CLIFFS QUICK REVIEW™

Plant Biology

By Patricia J. Rand, Ph.D.

CliffsQuickReview™ **Plant Biology**
Published by:
Course Hero, Inc.

Library of Congress Control Number: 00-103373
ISBN: 978-1-957671-25-3

10 9 8 7 6 5 4 3 2 1
1O/RX/QR/QR/IN

CONTENTS

CONTENTS

CONTENTS

CONTENTS

CONTENTS

CHAPTER 1
THEMES OF PLANT BIOLOGY

What Is a Plant?

Although it may seem unnecessary to begin by defining "plant," in fact hundreds of researchers—including several Nobel Prize winners—
in laboratories all over the world are discovering previously unknown relationships among living things by examining for the first time the genetic codes that direct the very essence of being. In the process, our ideas on what constitutes a plant are changing. While trees are still obviously plants, and cats and dogs are still animals, the newly forming classifications separate algae and fungi (mushrooms) from the plant kingdom and give super-kingdom status to the bacteria. Thus restricted, the plant kingdom now includes, in general: mosses, liverworts, hornworts, ferns, fern allies, gymnosperms, and flowering plants, all of which are discussed later in this book (and whose characteristics are summarized in the next chapter). Instructors in most plant biology courses continue to discuss many of the removed "non-plants" because of the significance of these organisms to the origin and development of the acknowledged plants. Brief descriptions of some of the non-plants, accordingly, are also included in this book.

Characteristics of organisms

All living things, despite differences in appearance and size, share basic characteristics. Organisms:

- Are composed of **cells,** the smallest units able to conduct the functions of living.
- Have **genes,** sequences of **deoxyribonucleic acid** *(***DNA***)* that carry the instructions for the organization and functioning of the organism.

- Are made principally of four elements—**carbon, hydrogen, oxygen, nitrogen**—which were most abundant when the first life appeared eons ago on an early Earth. They combine to form the familiar compounds associated with life, such as—water (H_2O), carbon dioxide (CO_2), methane (CH_4), ammonia (NH_3) and a host of others.
- Need **energy** to conduct their **metabolism** (all of the chemical processes occurring within their bodies).
- Require **materials** from the environment to both build and maintain their bodies.
- Are structurally **organized.** Multicellular organisms build **tissues** (groups of similar cells that perform certain functions) and **organs** (structures formed of different tissues that act as a group to perform specialized functions).
- **React** to stimuli and **respond,** thereby **adapting** to their environment.
- **Grow** (increase in size or weight).
- **Reproduce,** producing offspring that insure the continuity of the genetic code from generation to generation.
- **Evolve** (change over time).

Special characteristics of plants
A plant has all the features of organisms listed above and, in addition most plants have the following **special plant characteristics:**

- Plants can **photosynthesize** (capture light energy and make organic compounds from inorganic materials), which makes them different but not unique—a few other organisms also are photosynthetic, such as some algae and bacteria.
- In the **life cycle** of plants there is an **alternation of generations** in which two genetically different plant bodies alternate: a haploid **gametophyte** alternates with a diploid **sporophyte.**

- Plants develop from **embryos,** immature sporophytes formed by a fusion of egg and sperm cells, supported by nonreproductive gametophytic tissue.
- Plants have **indeterminate growth.** While animals reach a certain size and stop growing, plant cells in their **meristematic tissues** retain the ability to divide and grow throughout the life of the plant.
- Plants are **sedentary**, unlike most animals, but have evolved myriad ways to obtain the materials they need for their metabolism and efficient ways to reproduce and distribute their genes while anchored in one place.
- Although **lacking the nervous systems** of animals, plants react and adapt to environmental stimuli (with dramatic and surprising speed in some instances); they also produce **secondary metabolites,** chemical compounds not directly needed for survival, which deter other plants, fungi, and animals from attacking or consuming the plants.
- The terrestrial plants of today have evolved with a **dependence on water** (inherited from their aquatic ancestors); they have developed an elaborate system for obtaining, moving, using, and retaining water for all their metabolic processes and reproductive needs.

A phylogenetic tree of life

Biologists are interested in how organisms are related to one another and have as a general goal the construction of a **tree of life** in which the evolutionary relationships of all organisms are traced through time much as genealogists trace human family histories. In biology the study of developmental history and evolutionary relationships is called **phylogeny.** It is possible to trace *phylogenies* because

- Organisms have **heredity;** parents transmit genetic information to their offspring through the renowned molecule, **DNA** (**deoxyribonucleic acid**).

- Organisms change over time; they **evolve** to meet changing environmental conditions. The changes gradually become encoded in the DNA and separate **lineages** of organisms appear.

The sequence of base pairs in individual molecules of DNA and RNA are used to track relationships. Organisms with similar sequences are assumed to have had a common ancestor, and the more alike the strings of base pairs, the closer is the relationship. By combining molecular data with the information already known about organisms, biologists construct phylogenetic trees like that shown in Figure 1-1. Most, but not all, biologists agree that at present this is the best way to arrange the branches, but new data undoubtedly will require new interpretations and, perhaps, different phylogenetic trees.

The Flowering Plants as "Typical" Plants

When plants are mentioned, most people visualize one of the large dominant plants of their region—perhaps a cactus for desert dwellers, or a vista of waving grasses in the prairie, or tall sycamores along a river in the Midwest. All of these plants are **angiosperms** or **flowering plants.** Except for the trees of the coniferous forests, most of the large, visible plants around us in the temperate zone and the tropics are angiosperms. In past geologic eras they did not form the dominant vegetation During earlier eras, **gymnosperms, ferns,** or **fern allies** were the principal players on Earth, and before a terrestrial flora appeared, **bacteria, algae,** and **protista** colonized the primeval waters.

There are close to 300,000 described species of angiosperms in the world. They are the plants we rely on for our food and shelter. They are feed for livestock and provide us with aesthetically pleasing landscapes and gardens. They make all life possible by capturing the energy of the sun and transforming it into a form usable by animals.

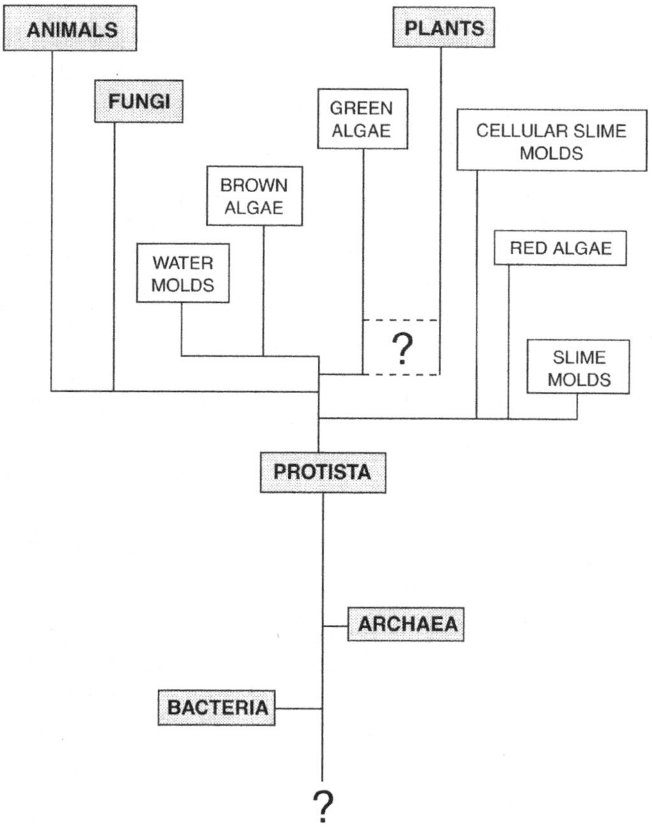

Figure 1-1

Angiosperms are studied in thousands of laboratories and fields and probably more is known about their characteristics than those of any other plant group.

Therefore in this book and most introductory plant biology texts, the term "plant," unless designated otherwise, generally refers to an angiosperm and "typical" plants are flowering plants. Note, however, that while most plants share many of the same types of tissues, cells, and functions, there is no *typical* plant with *all* of the described characteristics; in the real world there probably are exceptions to almost all of the textbook generalities made about "typical" plants.

Flowering plants have **stems, leaves,** and **roots** and reproduce by **seeds** located in **fruits** produced in **flowers.** As most of these structures are distinctive, neither a microscope nor the help of a botanist is needed usually to identify the various parts. Study of the details of plant structure is an important part of plant biology because plant function is integrated inextricably with form and structure. To know one helps in understanding the other. A knowledge of structure leads to a greater appreciation of how well-designed plants are for carrying out their functions.

Divisions of Plant Science

Biology, the study of life, historically was divided into **zoology,** the scientific study of animals, and **botany,** the scientific study of plants, which in the past included the **bacteria, algae, lichens, fungi.** Botany today is more often called **plant biology** or sometimes **plant science.**

Botanists may specialize in a subject matter field like **ecology** or **genetics** or else study all aspects of one group of plants, (and are then called **bryologists** if they study mosses, for example, or **pteridologists** if they study ferns). The plant scientists in applied fields like **forestry** or **horticulture** rarely call themselves botanists, but they often study the same processes and structures as do botanists. **Microbiologists** investigate microorganisms such as bacteria and may specialize in such aspects as **microbial ecology. Molecular biologists** are interested in the structure and function of biological macromolecules and study such processes as the biochemical aspects of genetics while **plant physiologists** analyze metabolic processes such as photosynthesis.

The Scientific Method

Botany, or any other science, is a body of knowledge accumulated over time by its practioneers using a process called the **scientific method.** Both the facts and the process used to obtain them are important in understanding botany. The scientific method varies in details and application by its users, but in general consists of the following steps: A problem is defined; information is accumulated; a hypothesis (proposed answer) is formulated and tested experimentally; if the new data contradict the hypothesis, the hypothesis is revised and tested until a conclusion is reached that explains the phenomenon of interest. The final step is to publish the results so other interested scientists will design experiments to validate or refute the work.

Science is an ever-changing quest; new hypotheses are tested daily, and new information and conclusions are added constantly to the data base. When a sufficiently large number of experiments reach the same conclusions, the hypothesis is incorporated with others and becomes a **theory.** Theories are the foundations of scientific knowledge and give rise to **principles** or **laws.** While theories sometimes are modified or changed as new evidence accumulates, principles are rarely, if ever, altered.

CHAPTER 2
A SYNOPSIS OF THE LIVING WORLD

Major Groups of Organisms

Faced with the multiplicity of organisms in the world, classifying them into groups with like characteristics makes talking about them, identifying them, and describing them easier. Utilitarian groupings—poisonous plants, trees that produce edible fruits, animals that sting, and so forth—are useful today as in the past, however, they provide information about only one aspect of organisms' lives. For decades researchers have accumulated libraries of information about creatures large and small. **Systematists** (the biologists concerned with classification and naming of organisms) have arranged much of this information into schemes that provide not only descriptions and names for organisms, but also show hereditary relationships among and between members of the various groups. Indeed, one of the principal objectives of **systematics** is to determine as best as possible—using all available information—the evolutionary history of life on Earth.

Not everyone agrees on how to organize and classify the existing information, and with researchers constantly producing new data, opinions change often to accommodate the new material. Authors of current textbooks reflect the differences of opinion. Which is the correct classification? Obviously, *the one used by your instructor.* This classification scheme, or any of the others, is neither completely right nor completely wrong. Any classification scheme is a **hypothesis** proposed by knowledgeable experts and subject to change and interpretation by other knowledgeable experts. Tomorrow's classifications will be different and probably closer to the truth than are today's.

The first column in Table 2-1 shows the 250-year-old Linnaean view of the living world. For centuries, if something was alive, it was either a "plant" or an "animal", or, later, a "higher" animal (such as a human being). When microscopes revealed a new world of tiny organisms living in concert with the more familiar larger ones, classifiers added

Table 2-1: Three Historical Classification Schemes of Organisms

Linnaean 2 Kingdom View	Century-Old Four-Division View
Plantae: bacteria, algae, fungi, slime, molds, lichens, mosses, liverworts, ferns, fern allies, cycads, ginkgos, conifers, flowering plants	Kingdom **Plantae** Division **Thallophyta** Subdivision **Algae**
Animalia: protozoa, sponges, multicellular animals	Subdivision **Fungi** - bacteria, yeasts, molds, mushrooms, slime molds, lichens
	Division **Bryophyta** Class **Musci**: mosses Class **Hepaticae**: liverworts Class **Anthocerotae**: hornworts
	Division **Pteridophyta** Class **Lycopodineae**: club mosses Class **Equisetineae**: horsetail Class **Filicineae**: ferns
	Division **Spermatophyta** Subdivision **Gymnospermae**: conifers Subdivision **Angiospermae**: flowering plants Subdivision **Dicotyledoneae**: roses, buttercups, beans, etc. Class **Monocotyledoneae**: grasses, etc.

Table 2-1: Three Historical Classification Schemes of Organisms

50-Year-Old Phylum View

Kingdom **Plantae**
- Subkingdom **Thallophyta**
  - Phylum **Cyanophyta**: blue-green algae
  - Phylum **Euglenophyta**: euglenoids
  - Phylum **Chlorophyta**: green algae
  - Phylum **Chrysophyta**: yellow-green algae, diatoms
  - Phylum **Pyrrophyta**: dinoflagellates, cryptomonads
  - Phylum **Phaeophyta**: brown algae
  - Phylum **Rhodophyta**: red algae

  - Phylum **Schizomycophyta**: bacteria
  - Phylum **Eumycophyta**: true fungi
- Subkingdom **Embryophyta**
  - Phylum **Bryophyta**
    - Class **Musci**: mosses
    - Class **Hepaticae**: liverworts
    - Class **Anthocerotae**: hornworts
  - Phylum **Tracheophyta**: vascular plants
    - Subphylum **Psilopsida**: whisk ferns
    - Subphylum **Lycopsida**: club mosses
    - Subphylum **Sphenopsida**: horsetails
    - Subphylum **Pteropsida**
      - Class **Filicineae**: ferns

      - Class **Gymnospermae**: conifers and allies
      - Class **Angiospermae**: flowering plants
        - Subclass **Dicotyledoneae**: dicots
        - Subclass **Angiospermae**: monocots

and refined categories—and have been doing so ever since to accommodate large and small discoveries.

A Current Classification of Plants and Other Organisms

A century ago, botanists recognized four major groups of plants, but within 50 years these groups had been subdivided and rearranged into still further groupings. Table 2-1 presents the old groupings of plants in considerable detail because some of the names persist as common names for modern plant groups, for example, thallophytes, gymnosperms, and monocots. By and large, the major groups (such as ferns, mosses, and flowering plants) are still recognized as units in today's more recent rearrangements but now appear with new technical names.

A current lively debate among plant systematists concerns how best to incorporate the recently obtained data from molecular biology into a phylogenetic arrangement of the plant kingdom. (Configurations in the past were based primarily upon morphological and anatomical features.) Of interest as well is the controversy over the ranking of the two clearly different groups of bacteria. Are they "kingdoms," "superkingdoms," or "domains"? Should the classification of all living things start by first separating the prokaryotes and eukaryotes into two superior groups? Table 2-2 summarizes three current ways. Note that the four eukaryote groups are retained as "kingdoms" in all of the classifications.

Most plant biology texts (including this one) mention briefly the multicellular animals and discuss in detail only the most plant-like of the protista. Bacteria and fungi are customarily included, although scientists no longer considered them to be plants. Table 2-3 gives an overview of the major groups usually studied in introductory plant biology courses. As always, your textbook is the best place to find pictures, more detailed information about plants and definitions of unfamiliar terms.

Table 2-2: Three Ways to Classify Organisms

3 Domains, 4 Kingdoms	*6 Kingdoms*	*2 Superkingdoms, 5 Kingdoms*
		Superkingdon **Prokarya**
		Kingdom **Bacteria** (Prokaryotae, Procaryotae, Monera)
Domain **Archaebacteria** (Archaea): bacteria of extreme environments	Kingdon **Archaebacteria**	Subkingdom **Archaea**
Domain **Bacteria**: most of the common bacteria	Kingdom **Eubacteria**	Subkingdom **Eubacteria**
Domain **Eukarya**		Superkingdom **Eukarya**
Kingdom **Fungi:** molds, rusts, yeasts, smuts, morels, truffles, mushrooms, lichens	Kingdom **Fungi:** molds, rusts, yeasts, smuts, morels, truffles, mushrooms, lichens	Kingdom **Fungi:** molds, rusts, yeasts, smuts, morels, truffles, mushrooms, lichens
Kingdom **Protoctista (Protista):** algae, slime molds, water molds, protozoans	Kingdom **Protoctista:** algae, slime molds, water molds, protozoans	Kingdom **Protoctista:** algae, slime molds, water molds, protozoans
Kingdom **Plantae:** mosses, liverworts, ferns, fern allies, cycads, ginkgos, conifers, flowering plants	Kingdom **Plantae:** mosses, liverworts, ferns, fern allies, cycads, ginkgos, conifers, flowering plants	Kingdom **Plantae:** mosses, liverworts, ferns, fern allies, cycads, ginkgos, conifers, flowering plants
Kingdom **Animalia:** multicellular animals	Kingdom **Animalia:** multicellular animals	Kingdom **Animalia:** multicellular animals

Table 2-3: A Synopsis of the Organisms Usually Studied in Introductory Plant Biology

Name	*Distinctive Features*	*History and Phylogeny*	*Other Significant Features*
Eubacteria (Bacteria): true bacteria, cyanobacteria, spirochetes, purple and green bacteria, pathogens	Prokaryotes; the most metabolically diverse—organisms: autotrophs (photosynthetic and chemosynthetic), hetertophs; anaerobes, and aerobes	The most abundant, smallest, and oldest organisms; present 2 billion years before eukaryotes appeared; modified the environment and made possible eukarryotic life	Recycle organic matter; fix atmospheric nitrogen; significant cause of diseases; used in industry to make cheese, alcohol, antibiotics, and in genetic engineering
Archaebacteria: extremophiles, methanogens, halophiles, thermophiles	Prokaryotes of extreme environments, i.e. very: hot, cold, acid, salty; structurally different from eubacteria and more closely related to the eukaryotes; appeared later than eubacteria	The endosymbiosis theory suggests some prokaryotes were engulfed by others and lived symbiotically within them; over time, these became the organelles of eukaryote cells	Extremophiles responsible for: natural gas reserves, colors in hot-springs, salt flats; live in hydrothermal vents on the ocean floor in high pressures and temperatures
Viruses (Not Organisms): bacteriophage, retrovirus, HIV, polio virus, tobacco mosaic cell; virus, rhinovirus, Ebola virus	Non-cellular, not alive; are molecules of DNA or RNA surrounded by like their own, need specific protein; need energy from a host cell to replicate; non-motile; don't grow nor metabolize	No fossil record; may have caused malformations seen in some fossils of early organisms; probably arose as bits of DNA broken from genomes of cells	Attack hosts with genomes most protein receptors to enter a presence triggers response, i.e. disease in plants, animals, bacteria, protists
Fungi: mushrooms, truffles, rusts, smuts, molds, yeasts, rots. Divided into four or five groups on basis of their sexual reproductive structures: Zygomycota, Ascomycota, Basidiomycota,	Not plants; eukaryotic body called mycelium consists of masses of hyphae (filaments); cell walls of chitin; glycogen stored as reserve food; are heterotrophic; obtain food by spores and sexually by zygotic meiosis	Oldest filaments appear in a Lower Cambrian; also some mycorrhizae in stems of an early Devonian plant; mycorrhizae may have facilitated the move of plants from water to land—with fungi	The principal decomposers together with the bacteria; most are saprotrophs (saprobes); some are causal agents of plant, animal, human diseases; many form important symbioses with roots of vascular plants

Deuteromycetes, Fungi Imperfecti. Chytridiomycota often included with Fungi		substituting for roots of the first land plants	(mycorrhizae); others are symbiotic with algae forming *lichens*; some edible, others used commercially in brewing, baking, medicine
Protista (Protoctista): algae, amoebae, flagellates, sporozoans, ciliates, water molds, diatoms, slime molds, etc. No completely satisfactory way to classify this extremely diverse group	Not plants; eukaryotes many with plant-like features (chlorophyll, photosynthetic, cellulose in walls), others like fungi (filamentous), some animal-like (ingest food); unicellular, multicellular, colonial; mostly aquatic, both marine and freshwater	Oldest eukaryote may be a brown-alga-like fossil 1.7 billion years old; or another presumed photosynthetic eukaryote found in 2.1-billion-year-old rocks; acritarchs (fossilized shells of shelled amoebas) first found in rocks 1.5 billion years old	Importantance: plants, animals, and fungi all derived from ancient protists; plant ancestor a green alga very much like green algae of today
Plantae– Bryophytes: mosses, liverworts, hornworts. Divided into three groups: Bryophyta (mosses), Hepaticophyta (liverworts), Anthocerotophyta (hornworts)	Small, nonvascular plants; body a thallus; gametophyte is free-living and the prominent plant; the sporophyte small, dependent upon the gametophyte; both sexual and asexual reproduction present; motile (flagellated) sperm require water in which to reach the egg	Ancestors of bryophytes probably derived from a green alga ancestor; oldest fossil bryophytes in 350 million years old rocks (younger than first vascular plant fossils, but probably because bryophytes lack resistant tissues for preservation)	Three groups are of different lineages, liverworts the oldest; mosses important in the ecology of the arctic and subarctic; some commercial use—peat for fuel, *Sphagnum* as packing material
Plantae–Seedless Vascular Plants: ferns, fern allies, and horsetails. Divided into four groups: Psilotophyta (whisk ferns), Lycophyta (club	Xylem and phloem present; sporophyte dominant; asexual spores produced in sporangia; some taxa homosporous, others heterosporous	Earliest vascular plants were small, simple, and dominated mid-Silurian to mid-Devonian landscapes (425–370 million years ago); by late Devonian	Most living taxa are depauperate remnants of large tree species of the Coal Age flora

(continued)

Table 2-3: A Synopsis of the Organisms Usually Studied in Introductory Plant Biology *(continued)*

Name	*Distinctive Features*	*History and Phylogeny*	*Other Significant Features*
mosses), Sphenopsida (horsetails), Pterophyta (ferns)		through the Carboniferous (the Coal Age), plants formed swamp forests of large trees; major groups died out by end of the Permian (250 million years ago)	
Plantae–Gymnosperms: conifers, cycads, gnetophytes, pines, junipers, yews, sequoia; divided into four groups: Cycadophyta (cycads), Ginkgophyta (*Ginkgo*), Coniferophyta (conifers), and Gnetophyta (gnetophytes)	Ovules and seeds exposed and not covered by sporophyte tissue at time of pollination; wood lacks fibers and contains only tracheids (except in gnetophytes, which have vessels in addition); some have flagellated sperm carried in a pollen tube toovule; female gametophyte produces several archegonia and embryos, but usually only one embryo survives/ovule	Seed plants probably evolved from seed-bearing progymnosperms sometime during the Devonian (370 million years ago); current gymnosperms represent a series of separate lineages	Trees and shrubs with no herbaceous representatives (some vine-like *Gnetum* species are present); distributed world-wide from boreal conifer forests to tropics, some taxa mixed with Angiosperm trees, others in pure stands of Gymnosperms
Plantae–Anthophyta: angiosperms, flowering plants (see Chapter 23 for subdivisions)	Ovules enclosed in a carpel and seeds produced in fruits; both gemetophytes highly reduced (female a seven-celled structure, male the germinated pollen grain); sporophyte the dominant plant	First fossil flowers appear in Cretaceous beds 130 million years old; in next 35 million years (Upper Cretaceous), angiosperms spread and became dominant plants in the Northern Hemisphere and, in next 10 million years, in the Southern Hemisphere also	Most successful of all the plant groups; insect-pollinated flowers basic; seeds with embryos enclosed in a resistant seed coat enabled plants to be widely dispersed; physiological adaptations and production of secondary metabolites of importance in their success

CHAPTER 3
CELLS

Cell Theory

The modern **cell theory,** one of the fundamental generalizations of biology, holds that:

- All organisms are composed of one or more cells.
- New cells come from pre-existing cells; lifeforms today have descended in unbroken continuity from the first primitive cells that arose on earth more than 3.5 billion years ago.
- Hereditary information passes from parent cell to daughter cell.
- The fundamental biochemical reactions of life take place within cells.

Methods of studying cells

Microscopes. Cytology—the scientific study of cells—has progressed simultaneously with the development of better, more powerful microscopes and cell preparation techniques. **Light microscopes,** invented in the 1500s, today can magnify in the range of 100x to 1000x and are at the maximum of their resolving power at about 0.2 Φm (micrometer = $^1/_{1,000,000}$ meter). They are limited from further improvement by the wavelength of visible light that is used to create the image (the shorter the wavelength, the greater the resolution).

Beams of electrons have shorter wavelengths than visible light, hence **electron microscopes** (which focus streams of electrons) are able to achieve resolutions of 2–0.4 nm (nanometer = $^1/_{1,000,000,000}$ meter = 1/1000 Φm) and magnifications of over 100,000x. One type of electron microscope, the **transmission electron microscope (TEM),** focuses a beam of electrons *through* an object much as a light beam is used in light microscopes, whereas a **scanning electron microscope (SEM)** forms an image from electrons that are scattered from the *surface* of a specimen.

Laboratory techniques. In light microscopy dyes that stain particular chemicals are commonly used to identify cellular structures in thin sections of tissue mounted on microscope slides. Specimens observed by electron microscopy are cut in much thinner sections (electrons don't penetrate material well) and examined in a vacuum. These preparations cause problems of interpretation of the microscopic image. Does the material *really* look like this in living cells or has preparation altered its appearance? Living material can not be examined using electron microscopes so the answer is moot.

Today's understanding of cells is based on two approaches: 1.) the sometimes very difficult biochemical analyses of cells and 2.) the increasing greater magnification of their constituent parts.

Cell size

Plant cells vary in size from about 7 Φm—the diameter of dividing cells at the tips of roots and shoots—to fiber cells a few micrometers wide but a meter or more in length. Most plant (and animal) cells are between 10–100 Φm and thus too small to be seen without **magnification** since human eyes have a **resolving power (resolution)** of about 100 Φm. (Resolving power is the ability to see adjacent objects as separate; magnification simply makes things larger.

Two things closer together than 100 Φm will look like one to the human eye unless a microscope with a resolving power of less than 100 Φm is used.

Limits to cell growth

Plant cells are separated from their environment by a **cell wall** inside of which a semi-liquid **plasma membrane** surrounds the **cytoplasm** and the **nucleus.** As cells grow, their volume increases faster than their surface area, a geometric reality (volume increases as the cube of the diameter, surface as the square of the diameter in the **surface:volume ratio**). Since materials used and released in the metabolism of the cell must pass through its surface, a functional upper size limit to growth is reached; cells can grow no larger in volume than the membrane area can support. Distribute the same volume of cytoplasm contained in one large cell into several smaller cells and the surface area increases, the cell can function, and ultimately, **multicellular** organisms result. Multicellular bodies have advantages over **unicellular** ones, e.g. cells can specialize to perform specific biochemical functions, different kinds of structural materials and structures can be produced, and the control center of the cell, the nucleus, is more efficient with fewer activities to control.

Two kinds of cells: Prokaryotes and eukaryotes
All cells have

- An outer membrane, the **plasma membrane** (also called the **cell membrane** or **plasmalemma**), composed of a **lipid** bilayer in which **proteins** are embedded.
- Genetic material, found in molecules of **deoxyribonucleic acid (DNA).**

The fundamental separation. Based on cellular structure, the living world can be divided into two kinds of organisms: the **prokaryotes** and the **eukaryotes.** The cells of prokaryotes (the **bacteria**) lack a **nucleus** and other **membrane-bounded** cell **organelles** whereas those of the eukaryotes—the rest of the living world—have nuclei and a complex internal structure with membrane-bounded cell organelles. Table 3-1 compares features of the cells of the two basic types.

Table 3-1: Comparison of Prokaryote and Eukaryotes

Features	*Prokaryotes*	*Eukaryotes*
Average size of cells	1–10 Φm	10–100 Φm
Nucleus	absent (have a "nucleoid" region)	present as a distinct structure
Nuclear envelope (membrane)	absent	present, surrounds the nucleus
Genetic material (DNA)	present in one large circular molecule, which acts as a "chromosome" in the nucleoid region	present in linear chromosomes in the nucleus, usually many per cell
DNA bound to histones (proteins)	no	yes
Cytoskeleton of microtubules and actin filaments	absent	present
Membrane-bounded organelles	absent	present
Ribosomes	present	present
Plasma membrane	present	present
Cell wall	present	present in plants and, some other organisms

Origin of the eukaryote type. Biologists in general agree that early organisms were prokaryotes that developed ways to survive on a young Earth whose atmosphere and conditions were much different from today's. Descendants of these ancient prokaryotes—the organisms of today—carry in their genes solutions to old problems as well as modifications made over 3.5 billion years in response to changing Earth conditions.

Serial endosymbiotic theory

The mitochondria and chloroplasts of plants differ from other organelles in their semi-independence. In many ways they resemble bacteria: Both bacteria and the organellescan divide, their DNA is organized similarly, and both have small ribosomes. Additionally, antibiotics inhibit protein synthesis in ribosomes of mitochondria, chloroplasts, and bacteria, but have no effect on synthesis in cytosomal ribosomes of eukaryotic cells. These structural and physiological differences lead some biologists to postulate the **serial endosymbiotic theory** of organelles in eukaryote cells. (A **symbiosis** is the close association and living together of individuals of different species.) The theory suggests that, early in the development of life on Earth, what are now organelles of larger, eukaryotic cells were simple, free-living prokaryotes. At some point in time, each independently developed a symbiotic relationship with larger, heterotrophic cells bringing to the partnership energy-capturing and -converting abilities and receiving in trade a physical home that provided the chemical necessities for life and protection from severe environmental stresses. The combination proved successful. Cells with the combined attributes were able to colonize previously hostile environments and the stage was set for the invasion of land and development of green, multicellular, terrestrial plants. The process occurred repeatedly—it was a *serial endosymbiosis*.

The Generalized Plant Cell

Four features set apart plant cells from those of other organisms:

- A **cellulose cell wall;** many protists, some fungi, and most bacteria also have rigid walls, but made of different materials, e.g. Chitin strengthens fungal cell walls and peptidoglycan those of bacteria.
- **Plasmodesmata,** strands of cytoplasm that protrude through pores in the cell walls and connect the protoplasts of adjacent cells; these are avenues of material transport in plants.

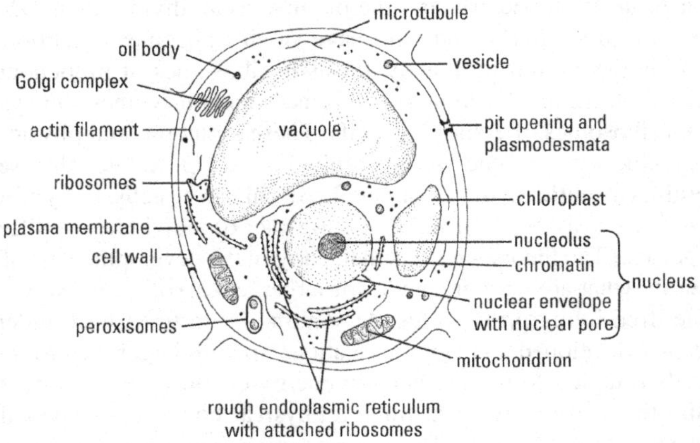

Figure 3-1

- **Plastids,** organelles thought to have an endosymbiotic origin; they have an intricate internal structure of folded membranes that greatly increases the internal surface area on which chemical reactions occur; **chloroplasts** contain **chlorophyll,** and their **thylakoid membranes** are the sites of **photosynthesis.**
- **Vacuoles** are fluid filled sacs present in plants, animals, and some protists; those of plants differ from others in size and function, i.e. a typical plant cell has a single, centrally located vacuole whose water-filled interior pushes the cytoplasm outward against the cell wall giving rigidity or turgidity to the cell.

Not all functional cells in plants are alive. Many cells live for only a short time then die, but their cell walls remain in place giving strength and rigidity to the plant body. The dead cells of **xylem** tissue form effective conduits for water between the roots and the above ground tissues.

Plant cells vary greatly in shape and function, but in general each has a **cell wall** surrounding a **protoplast,** which is differentiated into a **nucleus** and **cytoplasm**. Suspended within the watery cytoplasm are the membranes, organelles, and particles in which the biochemical processes of life occur. The **nucleus** contains most of the cellular DNA and is thus the center from which these cellular activities are directed. Cytoplasm is in more or less constant motion in living cells and the movement is termed **cyclosis** or **cytoplasmic streaming**. (**Protoplasm** is a commonly used name for all the watery cellular matter. It incudes the **nucleoplasm** of the nucleus plus the cytoplasm.)

The *cell wall* is produced by the cytoplasm and deposited outside of the **plasma membrane**—the outer boundary of the protoplast. The wall and plasma membrane regulate the kinds of materials that enter and leave the cell thus making it possible for the cell to maintain an identity different from its surroundings. The wall is not simply an inert substance providing strength and giving structure to plant cells as once was thought, but is active in several metabolic processes including **absorption** and **secretion** of substances, the **detection** of bacterial and fungal pathogens, and even **regulation of growth** and development.

Cellulose, the most common material of plant cell walls, is a polysaccharide composed of long chains of glucose molecules that assume crystalline properties in the cell wall. During wall formation, adjacent cellulose molecules link together forming bundles called **microfibrils.** The microfibrils, in turn, twist together rope-like, producing strong cords called **macrofibrils.** Hemicelluloses and pectins are deposited among the fibril network and chemically bind the whole together. The result is a **primary wall.** The whole process of synthesis and assembly is coordinated by the plasma membrane.

Lignin, the second most common material of plant cell walls, adds additional strength to cells, such as in wood cells. Other substances are deposited in walls of tissues serving particular functions. Walls of cells in the outer layer of leaves often contain **cutin** or **waxes,**

both of which effectively reduce the loss of water from the leaves. Walls of **cork** cells in the bark of trees contain **suberin,** another type of protective material.

Membranes:

- Are the sites of many of the biochemical processes of life.
- Are in constant flux, changing both their physical and chemical structure and composition as long as the cell is alive.
- Partition the cell into compartments in which metabolic reactions take place independently of processes occurring in nearby compartments, thus permitting a variety of reactions to occur simultaneously within a single cell.

The basic structure of all biological membranes is the same: a double layer of **lipids** in which **proteins** are embedded. The ratio of proteins to lipids is 3:2. Some of the proteins extend through the lipid layer, each end of a molecule projecting out opposite sides. Others remain on or partially embedded in one or the other surface.

Most membranes are **selectively (differentially) permeable,** permitting or preventing materials from leaving or entering the cell. When toxic substances destroy the selective permeability and the membrane becomes **freely permeable,** materials leak unchecked in and out, resulting in death of the cell.

Although it is difficult to study the plasma membrane separately from the rest of the protoplast, it is known that the plasma membrane deciphers the biochemical signals that control cell growth and differentiation and that it organizes the formation of the cell wall.

Table 3-2 shows cell structures, their composition, and principal functions.

Table 3-2: Cell Structure and Function

Structure	*Function*
Cell wall	Outer layer of plant cells; produced by the cytoplasm; gives shape and rigidity to the cell; cellulose the basic constituent
Middle lamella	Intercellular layer (mostly pectin) between primary walls of adjacent cells; binds them together
Primary wall	First wall deposited by actively growing and dividing cells
Secondary wall	Deposited inside primary wall after cell has stopped growing; cellulose, lignin, other materials deposited in layers give strength to plant; **pits** present—areas where no secondary wall is deposited and through which plasmodesmata extend
Plasmodesmata	Strands of cytoplasm that connect adjacent cells; are pathways for material movement
Protoplast	All of the material contained within the cell wall
Nucleus	Structure that contains the genetic information (DNA) in eukaryotic cells; controls cellular activities
Nuclear envelope	Pair of fused membranes around the nucleus; connected to the endoplasmic reticulum; contains pores through which the nucleoplasm and cytoplasm connect
Nucleoplasm	The fluid portion of the nucleus; also called the nuclear matrix
Chromatin	In non-dividing cells, threads of **deoxyribonucleic acid (DNA)** plus associated proteins **(histones)** that are attached to sites on the nuclear envelope; condenses into a compact mass when cells divide, forming **chromosomes** that carry the **genes**
Nucleolus	Plural: **nucleoli**; one or more spherical structures that are the site(s) of **ribosomal ribonucleic acid (rrna)** assembly from rrna genes; subunits of ribosomes also produced here
Cytoplasm	Living cellular material exclusive of the nucleus

(continued)

Table 3-2: Cell Structure and Function *(continued)*

Structure	*Function*
Plasma membrane	Outer boundary of the cytoplasm; a lipid bilayer with embedded proteins; **differentially permeable** and regulates movement of materials into and out of cells; coordinates synthesis of cell wall; recognizes and transmits internal and external chemical signals
Cytosol	Liquid portion of the cytoplasm in which cellular structures are suspended; also called the cytoplasmic matrix
Organelles	General name for cellular structures bounded by membranes and specialized to perform specific functions
Bounded by two membranes	Evidence that these organelles once were independent prokaryotes; they retain many of their former traits within the eukaryote cells
Plastids	Semiautonomous, contain DNA and ribosomes and reproduce by fission; have an elaborate internal structure; in algae and plants
Chloroplasts	Green (contain **chlorophyll**) sites of photosynthesis in **thylakoid** membranes, amino acid and fatty acid synthesis
Chromoplasts	Yellow, orange, red (contain **carotenoid** pigments); attract pollinators to flowers and dispersers to colored seeds and fruits
Leucoplasts	Colorless site with no pigments; **amyloplasts** synthesize starch; others synthesize oils and probably proteins
Mitochondria	Singular: **mitochondrion**; sites of aerobic respiration and release of **adenosine triphosphate (ATP)**; similar tolike plastids in being semiautonomous and containing DNA and ribosomes; also reproduce by fission; inner membrane with many folds or **cristae**
Bounded by one membrane	Phospholipid bilayer that regulates material
Peroxisomes	Also called microbodies; no internal membranes, DNA, or ribosomes, but are self-replicating; some important in photorespiration; glyoxisomes contain enzymes that convert fats to sucrose during seed germination; others associated with mitochondria

Vacuoles	Sac of liquid, the **cell sap**, surrounded by a membrane, the **tonoplast**; in mature cells may occupy 90 percent of the cell; gives **turgor** (rigidity) to the cell; serves as temporary storage site for Calcium and other materials; **antho-cyanin** pigments in cell sap give color (reds and, blues) to leaves and flowers; some small vacuoles (like animal lysosomes) are sites of digestion; others store wastes
Endomembrane system	Phospholipid bilayer that regulates material; collective name for all the cell membranes except those of mitochondria, plastids, and peroxisomes; membranes originate in the ER
Endoplasmic reticulum (ER)	An extensive membranous system of flattened sacs (**cisternae**) that extends throughout the cytoplasm as a communication and transport system; **rough ER** is covered with ribosomes and delivers proteins; **smooth ER** lacks ribosomes,synthesizes lipids; rough ER is cisternal, smooth ER is tubular
Golgi complex	Collection of **Golgi bodies (dictyosomes)** that are stacks of flattened cisternae associated with secretion; some synthesize and export polysaccharides; others handle glycoproteins
Vesicles	Small sacs of secretory material pinched off from the cisternae; move from the Golgi complex to the plasma membrane (with the assistance of actin filaments) and liberate their contents outside of the cell; process is called **exocytosis**—secretion of materials carried in vesicles from the cell
Cytoskeleton	Matrix of protein fibers that gives support and on which organelles, enzymes, macromolecules are attached; composed of two kinds of protein filaments with similar functions
Microtubules	Long hollow tubes composed of the protein **tubulin**; in constant flux, breaking down and reforming; function in cell division, cellular movement, and movement of materials within the cell
Actin filaments	Long chains of the protein **actin**; responsible for cytoplasmic streaming, movement of nucleus in cell division, organization of the ER, and other movements of cellular materials
Ribosomes	Sites of protein assembly in the cytoplasm or on the rough ER; are small (17-23 Φm) particles assembled from a **large** and a **small subunit** produced in the nucleolus; are half **ribosomal RNA (rrna)** and half proteins in composition; **messenger RNA (mrna)** brings code from a gene, attaches to rrna and initiates protein synthesis; at sites of active synthesis clusters of ribosomes are called **polysomes** or **polyribosomes**

(continued)

Table 3-2: Cell Structure and Function *(continued)*

Structure	*Function*
Oil bodies	Spherical drops of lipids (especially triglycerides used to synthesize membranes) common in the cytoplasm of cells of seeds and fruits; used as energy and carbon sources for developing seedlings; not bound by a membrane; synthesized in plastids or in the ER
Flagella and cilia	Singular: flagellum and cilium; extensions of cytoplasm enclosed by the plasma membrane that project from the cell wall; made of two microtubules surrounded by nine others (a **9 + 2 structure**); cilia same structure but shorter; are used as locomotor structures by algae and protists; the only flagellated cells in plants are the motile sperm of mosses, liverworts, ferns, cycads, and ginkgo

CHAPTER 4
TISSUES

Plant Body

All but a relatively few plants are **multicellular,** and the majority have **bodies** comprised of two major systems: the **root system** and the **shoot system.** The former is usually underground, and the latter above ground. To succeed and grow simultaneously in two such entirely different environments—air and soil—requires a myriad of adaptations, starting with cellular modifications into specialized kinds of **tissues** (groups of similar cells that are organized in a structural and functional unit) followed by development of **organs** (structures composed of several kinds of tissues grouped in a structural and functional unit). The acquisition of form and structure is called **morphogenesis** and is a highly orchestrated procedure controlled by the DNA of the plant cells but influenced as well by the environment.

Growth and Development

"Development" and "growth" are sometimes used interchangeably in conversation, but in a botanical sense they describe separate events in the organization of the mature plant body.

Development is the progression from earlier to later stages in maturation, e.g. a fertilized egg *develops* into a mature tree. It is the process whereby tissues, organs, and whole plants are produced. It involves: **growth, morphogenesis** (the acquisition of form and structure), and **differentiation.** The interactions of the environment and the genetic instructions inherited by the cells determine how the plant develops.

Growth is the irreversible change in size of cells and plant organs due to both **cell division** *and* **enlargement.** Enlargement necessitates a change in the elasticity of the cell walls together with an increase in the size and water content of the vacuole. Growth can be **determinate**—when an organ or part or whole organism reaches a certain size and then stops growing—or **indeterminate**—when cells continue to divide indefinitely. Plants in general have indeterminate growth.

Differentiation is the process in which generalized cells specialize into the morphologically and physiologically different cells described in Table 4-1. Since all of the cells produced by division in the meristems have the same genetic make up, differentiation is a function of which particular genes are either expressed orrepressed. The kind of cell that ultimately develops also is a result of its location: Root cells don't form in developing flowers, for example, nor do petals form on roots.

Mature plant cells can be stimulated under certain conditions to divide and differentiate again, i.e. to **dedifferentiate.** This happens when tissues are wounded, as when branches break or leaves are damaged by insects. The plant repairs itself by *dedifferentiating* parenchyma cells in the vicinity of the wound, making cells like those injured or else physiologically similar cells.

Plants differ from animals in their manner of growth. As young animals mature, all parts of their bodies grow until they reach a genetically determined size for each species. Plant growth, on the other hand, continues throughout the life span of the plant and is restricted to certain **meristematic** tissue regions only. This continuous growth results in:

- Two general groups of tissues, **primary** and **secondary.**
- Two body types, **primary** and **secondary.**
- **Apical** and **lateral meristems.**

Apical meristems, or zones of cell division, occur in the tips of both roots and stems of all plants and are responsible for increases in the length of the primary plant body as the primary tissues differentiate from the meristems. As the vacuoles of the primary tissue cells enlarge, the stems and roots increase in girth until a maximum size (determined by the elasticity of their cell walls) is reached. The plant may continue to grow in length, but no longer does it grow in girth. Herbaceous plants with only primary tissues are thus limited to a relatively small size.

Woody plants, on the other hand, can grow to enormous size because of the strengthening and protective secondary tissues produced by lateral meristems, which develop around the periphery of their roots and stems. These tissues constitute the secondary plant body.

Meristematic Tissues

Meristematic tissues, or simply **meristems,** are tissues in which the cells remain forever young and divide actively throughout the life of the plant. When a meristematic cell divides in two, the new cell that remains in the meristem is called an **initial,** the other the **derivative.** As new cells are added by repeated **mitotic divisions** (see Chapter 14, Cell Division) of the initial cells, the derivatives are pushed farther away from the zone of active division. They stretch, enlarge and *differentiate* into other types of tissues as they mature. Meristematic cells are generally small and cuboidal with large nuclei, small vacuoles, and thin walls.

A plant has four kinds of meristems: the **apical meristem** and three kinds of lateral—**vascular cambium, cork cambium,** and **intercalary meristem.**

Apical meristems

These are located at opposite ends of the plant axis in the tips of roots and shoots. Cell divisions and subsequent cellular enlargement in these areas lengthen the above and below ground parts of the plant. The meristems also influence the shapes of the mature plants since the patterns for subsequent growth are laid down in the meristems.

Lateral meristems

Vascular cambium. Some plants grow in diameter by producing new tissues laterally from a cylinder of tissue called the **vascular cambium,** which extends throughout the length of the plant from the tips of the shoots to the tips of the roots. It is present in all **perennial** and in some **annual** plants. Tissues produced by cell divisions of the vascular cambium are **secondary tissues.**

Cork cambium. Cork cambia (singular: cambium), also called **phellogens,** are found in the bark of roots and stems of woody plants where they produce **cork** cells. The cork cambia originate just under the **epidermis** of the primary body and in some tree species are long cylinders running parallel to the vascular cambium. In other species, more discrete, disk-like cork cambia in the trunks produce flat plates of **bark** tissues that break off in large scales as the tree ages.

Intercalary meristem. Grasses have intercalary meristems located along the stems near the nodes. Cell divisions in this tissue push the stem upward. Grasses and other monocots have no lateral meristems so any lateral increase in size is the result of primary tissue cell enlargement, not cell divisions.

Primary (transitional) meristems

The cells produced by divisions in the apical meristem region are soon identifiable as three zones of distinct tissues that differentiate below the apical meristems. These are the **primary meristems,** called sometimes the **transitional meristems:** the **protoderm,** the **procambium,** and the **ground meristem.** They give rise to the tissue systems of the primary plant body.

Tissue Systems and Their Cellular Composition

The basic plant cell types, the tissue systems in which they occur, their location within plants, and a brief description of their principal functions are presented in Table 4-1.

Table 4-1: Tissue Systems, Tissues, Cell Types, and Their Locations and Functions

Tissue System	*Tissues*	*Cell Types Present*	*Cell Characteristics*	*Location in Plant*	*Function in Plant*
Ground tissue system	Parenchyma tissue	Parenchyma	many sided (14 common); thin primary walls, living at maturity	throughout the plant; most common type of cell and tissue	most metabolic processes; storage, wound healing, and regeneration
	Collenchyma tissue	Collenchyma	elongate; primary wall unevenly thickened (thicker in corners); living at maturity	in patches near outside of stems, along veins of leaves; example: "strings" in celery	support of young growing plant (primary plant body); flexible support for soft organs
	Sclerenchyma tissue	Sclereid	cuboidal, with thick secondary wall; either living or dead at maturity	throughout the plant; example: the gritty texture of pears	form hard layers of shells (as in peanuts) and pits of fruit (such as peaches); occur in small groups around wounds
		Fiber	long with lignified thick secondary wall; usually dead at maturity	associated with xylem and phloem; example: "strings" in leaves of grasses	support; storage
Dermal tissue system	Epidermis	Parenchyma, guard cells, trichomes (hairs),	specialized, e.g. open and close stomata; cutinized outer walls; alive at maturity; leaf epidermal cells transparent, without	outer layer of primary plant body, herbaceous plants; broken and lost in secondary body development	protection; usually a single layer of cells; root hairs are out-growth of epidermal cells
	Periderm	Parenchyma, cork cells, sclereids, cork cambium	living cork cambial cells produce heavily suberized cork cells that are dead at maturity	bark of woody plants; first layers beneath the epidermis, later layers deeper; many cork cambia, not a single cylinder like vascular cambium	protection for older stems and roots; replaces epidermis

(continued)

Table 4-1: Tissue Systems, Tissues, Cell Types, and Their Locations and Functions ***(continued)***

Tissue System	*Tissues*	*Cell Types Present*	*Cell Characteristics*	*Location in Plant*	*Function in Plant*
Vascular tissue system	Xylem	Vessel element	elongate, lignified secondary wall with pits; dead at maturity; end walls with perforations	throughout the plant; elements lined up end to end form a vessel in xylem	conduct water and minerals; secondary walls add strength and support to plant body; principal cell type of angiosperm xylem
		Tracheid	long, tapering with lignified walls; have pits, but no perforations; dead at maturity	in xylem throughout the plant	principal water and mineral conducting element in gymnosperms and seedless vascular plants; of secondary importance to vessels in angiosperms
		Parenchyma, fibers	thin walled, living parenchyma and elongate, dead fibers are accessory storage site	in xylem throughout the plant	parenchyma: storage, repair; fibers: strength and non-conducting support
	Phloem	Sieve-tube elements	elongated; primary wall only; sieve areas on end walls called a sieve plate; living at maturity but lack a nucleus	in phloem throughout the plant; elements lined up end to end form a sieve tube	conducts dissolved carbohydrates and other foods in angiosperms
		Companion cells	living with variable, usually elongated shape; primary wall only; connected by plasmodesmata to sieve-tube elements	in phloem throughout the plant; derived from same mother cell as sieve-tube element	apparently sends ATP and signal substances to the enucleate sieve-tube elements thus controlling cellular metabolism of the sieve tube elements in angiosperms

	Sieve cell	elongated and tapering, living at maturity; primary cell wall with sieve areas; lacks a nucleus; cytoplasmn with much tubular endoplasmic reticulum	in phloem throughout the plant	conducts dissolved carbohydrates and other foods in gymnosperms
	Albuminous cell	elongated with primary wall; living at maturity; connected to the sieve cell by many plasmodesmata	in phloem throughout the plant	apparently sends ATP and signal substances to the enucleate sieve-tube elements thus controlling the cellular metabolism of the sieve tube elements in gymnosperms

CHAPTER 5
ROOTS

A Generalized View

In most vascular plants, roots are underground structures that anchor the plant and provide a means to absorb the nutrients and water needed for growth of the plant body. New root tips grow continuously throughout the life of the plant and provide the surfaces through which most of the nutrients and water move. Roots are used as storage organs for the food materials produced by the shoots. The major functions of roots, thus, can be summarized simply as absorption, conduction, storage, and anchorage.

Root Zones

As cells are added to the tip by repeated cell divisions, a young root elongates and leaves behind cells that **differentiate** and become the primary roots of the plant. Four areas of the young root traditionally are recognized, but except for the terminal area, are not distinctly separate. Their descriptive names are only partially correct in describing the activities taking place in each area. These regions, starting at the tip and moving upwards towards the stem, are **the root cap, zone of active cell division, zone of cell elongation,** and **zone of maturation.**

The first two are compacted in the first centimeter or less of the axis with the latter two no more than 4–5 centimeters from the tip. *Only the root cap and the cell division regions actually move through the soil.* After cells start to elongate and mature, no further extension takes place, and the root is stationary for the rest of its life.

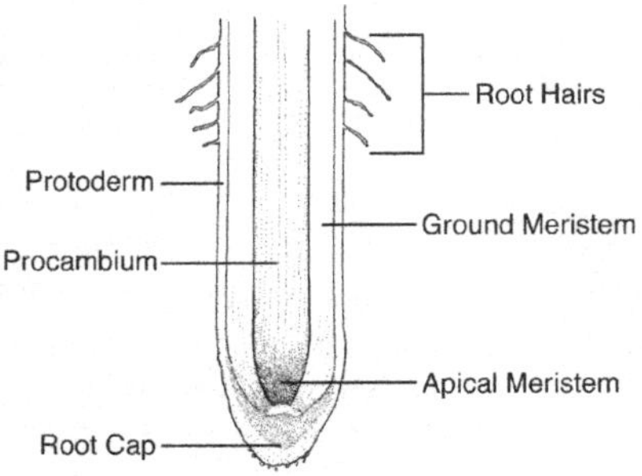

Figure 5-1

Root cap

The root cap is a cup-shaped, loosely cemented mass of parenchyma cells that covers the tip of the root. As cells are lost among the soil particles, new ones are added from the meristem behind the cap. The cap is a unique feature of roots; the tip of the stem has no such structure. From its shape, structure, and location, its primary function seems clear: It protects the cells under it from abrasion and assists the root in penetrating the soil. Phenomenal numbers of cap cells are produced to replace those worn off and lost as root tips push through the soil.

The movement is assisted by a slimy substance, **mucigel,** which is produced by cells of the root cap and epidermis. The mucigel

- Lubricates the roots.
- Contains materials that are inhibitory to roots of other species.
- Influences ion uptake.
- Attracts beneficial soil microorganisms.
- Glues soil particles to the roots thereby improving the soil-plant contact and facilitating water movement from the soil into the plant.

- Protects the root cells from drying out.

Root cap cells sense light in some as yet unexplained way and direct root growth away from light. The root cap also senses gravity to which roots respond by growing downward, bringing them into contact with the soil, the reservoir of nutrients and water used by plants. The root cap also responds to pressures exerted by the soil particles.

Zone of cell division

An **apical meristem** lies under and behind the root cap and, like the stem apical meristem, it produces the cells that give rise to the primary body of the plant. Unlike the stem meristem, it is not at the very tip of the root; it lies behind the root cap. Between the area of active division and the cap is an area where cells divide more slowly, the **quiescent center.** Most cell divisions occur along the edges of this center and give rise to columns of cells arranged parallel to the root axis. The parenchyma cells of the meristem are small, cuboidal, with dense protoplasts devoid of vacuoles and with relatively large nuclei.

The apical meristem of the root organizes to form the **three primary meristems**: **protoderm,** which gives rise to the epidermis; **procambium,** which produces xylem and phloem; and the **ground meristem,** which produces the cortex. Pith, present in most stems and produced from the ground meristem, is absent in most dicot (eudicot) roots, but is found in many monocot roots.

Zone of cell elongation

The cells in this zone stretch and lengthen as small vacuoles within the cytoplasm coalesce and fill with water. One or two large vacuoles occupy almost all of the cell volume in fully elongated cells. Cellular expansion in this zone is responsible for pushing the root cap and apical tip forward through the soil.

Zone of maturation

The elongating cells complete their differentiation into the tissues of the primary body in this zone. It is easily recognized because of the numerous **root hairs** that extend into the soil as outgrowths of single epidermal cells. They greatly increase the absorptive surface of roots during the growth period when large amounts of water and nutrients are needed. An individual root hair lives for only a day or two, but new ones form constantly nearer the tip as old ones die in the upper part of the zone.

Primary Root Tissues and Structure

The organization of tissues in the primary root is simpler than in the primary stem because no leaves are produced on the roots and, consequently, there is no need to connect the vascular system laterally to offshoots. The primary body, produced by the three primary meristems, consists of a central cylinder of vascular tissue, the **stele,** surrounded by large storage parenchyma cells—the **cortex**—on the outside of which lies a protective layer of cells—the **epidermis.**

Epidermis

The root hairs of the young epidermal cells vastly increase the surface area through which movement of materials can occur. The thread-like hairs are simply enlargements of the protoplast that extend outward into the soil. They have little wall material and are extremely fragile and easily broken. The root epidermis of some plants is covered by a thin, waxy **cuticle,** which apparently isn't thick enough to impede movement of substances through the epidermis.

Cortex

The cortex, composed primarily of parenchyma cells, is the largest part of the primary root, but in most dicots (eudicots) and in gymnosperms that undergo extensive secondary growth, it is soon crushed, and its storage function assumed by other tissues. Three layers of cortex are recognized: the **hypodermis** (also called exodermis), the **endodermis** and, between them, the **storage parenchyma.** The outer and inner layers of the cortex, the hypodermis and endodermis, are cylinders of tightly packed cells with heavily **suberized** walls and no intercellular spaces. (Suberin is the fatty substance that gives cork its distinctive attributes.) In contrast, the storage parenchyma cells are thin-walled and loosely packed with many intercellular spaces among them.

Hypodermis (exodermis). Just under the epidermis forming the outermost layer of the cortex is a layer one or two cells in width called the **hypodermis.** Since its cell walls are heavily suberized and impermeable to water its apparent function is to keep the water and nutrients (which are absorbed in the root zone further down the root) from leaking out through the cortex. The hypodermis is especially well developed in plants of arid regions and in those with shallow root systems. It also deters the entrance of soil microorganisms.

Endodermis. The innermost layer of the cortex is the **endodermis,** which is readily identifiable by the presence of **Casparian strips,** bands of suberin present on transverse and radial walls of its cells—the walls perpendicular to the surface of the root. The endodermis regulates the passage of water and dissolved substances by forcing them to move through living plasma membranes and plasmodesmata and not simply diffuse through the porous cell walls. The absorption and translocation of materials is thus selective; not everything in the surrounding soil gets through and into the plant body. An endodermis almost always is present in roots and generally never in stems.

Storage parenchyma. The bulk of the cortex consists of thin-walled, living parenchyma cells, which store **starch** and other substances. The cells expand or shrink as materials move in and out of their protoplasts. The large volume of air present in the intercellular spaces of this tissue provides important aeration for roots.

Stele (vascular cylinder)
The stele includes all of the tissues inside of the cortex: the pericycle, the vascular tissues—xylem and phloem—and, in some plants, a pith. Most dicot (eudicot) roots have a solid core of xylem in their center whereas most monocots have a pith composed of parenchyma.

Pericycle. The pericycle is a cylinder of parenchyma, one or at most a few cells in width, which lies in the stele immediately inside the endodermis. The cells retain their ability to divide throughout their lives, and localized divisions in the pericycle give rise to lateral (branch) roots. When secondary growth occurs in roots, the vascular cambium and usually the first cork cambium originate in the pericycle. Other cell divisions in the pericycle produce additional pericycle cells.

Vascular tissues. Most dicot (eudicot) roots differ from eudicot stems in having a lobed column of primary xylem as their core with phloem tissue occurring as strings of cells between the lobes. This arrangement is called a **protostele.** The primary xylem of monocots, on the other hand, forms a cylinder around a central mass of pith parenchyma, a **siphonostele.** The way in which the vascular tissues develop is useful in tracing ancestral relationships in the plant kingdom.

Secondary Growth of Roots

Secondary tissues comprise the greatest volume of the root mass of woody perennial plants. Primary tissues continue to form in the feeder roots, but the supporting root structure consists of **secondary tissues** produced by the lateral meristems, the **vascular cambium,** and one or more **cork cambia.** The usually unobserved underground root systems of most trees are as massive as the huge aerial bodies and counterbalance the aboveground weight thus keeping the tree upright and stable.

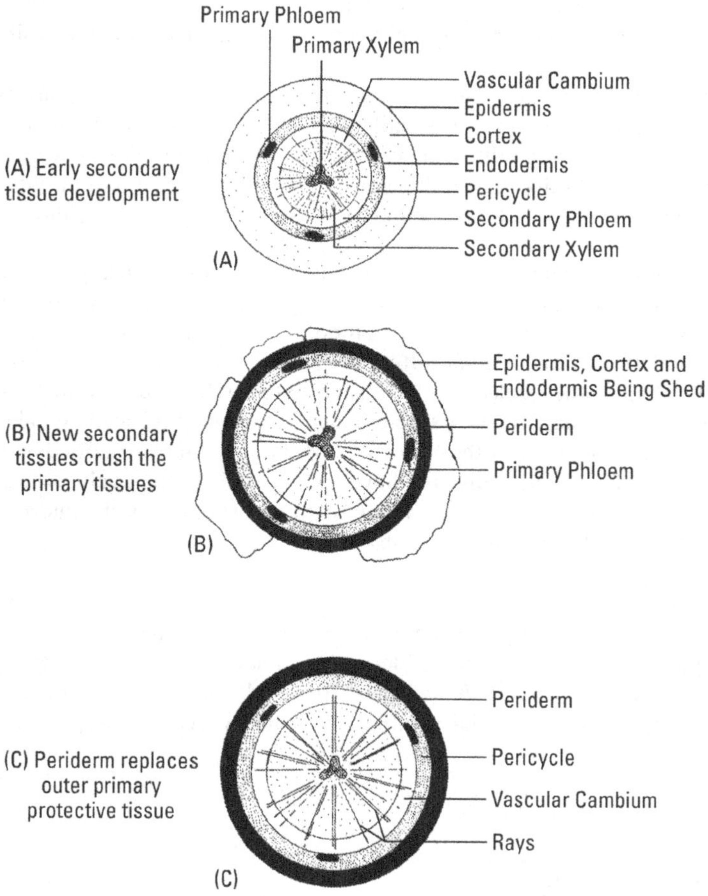

Figure 5-2

Roots produce branch roots and secondary tissues at the expense of the primary tissues. Cells in the primary tissue are broken and discarded as secondary growth proceeds. New lateral roots form

endogenously (from within the root) and push outward from the pericycle, destroying cortex and epidermal tissues on their way to the soil.

Initiation of secondary growth takes place in the zone of maturation soon after the cells stop elongating there. The vascular cambium differentiates between the primary xylem and phloem in this zone and pericycle cells divide simultaneously with the procambium initials. The result is a cylinder of cambium encircling the primary xylem.

The vascular cambium almost immediately begins producing xylem cells inward and phloem cells toward the outside of the root, in the process flattening the primary phloem against the more resistant endodermis. Concomitant differentiation of cork cambia in the pericycle adds other areas of cell division in the stele. The combination of periderm and vascular tissue production not only physically breaks the remaining cells of the cortex and epidermis, but the lignified and suberized new cell walls laid down by the cambia effectively isolate the outer tissues as well from their source of supplies in the interior of the root. Their death is inevitable.

By the end of the first year, secondary growth has obliterated all but the central core of primary xylem cells and a few fibers of primary xylem pushed against the periderm. The zones at this time, therefore, from outside to inside are: periderm, pericycle, primary phloem, secondary phloem, vascular cambium, secondary xylem, and primary xylem.

Types of Root Systems

Plants have three types of root systems: 1.) **taproot,** with a main taproot that is larger and grows faster than the branch roots; 2.) **fibrous,** with all roots about the same size; 3.) **adventitious,** roots that form on any plant part other than the roots. Fibrous systems are characteristic of grasses and are shallower than the taproot systems found on most eudicots and many gymnosperms.

Specialized and Modified Roots

Roots often perform functions other than support and absorption. Some store starch (beets and turnips) or water (desert plants). **Pneumatophores** are roots that grow into the air and are filled with a specialized parenchyma called **aerenchyma.** The large, intercellular spaces of aerenchyma are filled with oxygen and other gases. The pneumatophores seem to assist in aerobic respiration and gas exchange and are abundant on woody plants like cypress and mangrove, which grow in water-logged soils.

Some roots produce **suckers** from adventitious bud-like growths. Suckers grow into aerial shoots capable of independent existence, and serve to propagate the plant. Because they have the same genetic make-up as the parent plant, they are **clones** of the parent. Aspen groves in the western United States frequently are clones of only one or a few parent trees.

Many flowering plants develop as parasites on other plants. They produce adventitious roots called **haustoria** that penetrate the tissues of the host and connect to the vascular system, thus becoming part of the host pipelines. Parasitic roots lack most of the tissues of ordinary roots. Dodder and mistletoe are two of the more than 3000 flowering plant parasites.

Mycorrhizal roots are known from 90 percent of plant species and are a mutualistic association of a fungus with plant root tissue. Most plants *require* specific mycorrhizal fungi without which they are unable to absorb sufficient quantities of P, Zn, and Mn. The fungus takes the place of root hairs and may penetrate the cortex completely (**endomycorrhizae**) or remain on the surface of the root (**ectomycorrhizae**).

CHAPTER 6
THE SHOOT SYSTEM: STEMS

Characteristics of Shoot Systems

The aboveground, conspicuous part of flowering plants constitutes the **shoot system,** which is composed of erect **stems** on which are attached **leaves, flowers,** and **buds.** Leaves are attached to the **stem** at regions called **nodes**. The section of stem between nodes is an **internode,** and the upper angle between the stem and the leaf at the node is called the **leaf axil. Axillary (lateral) buds** located in the leaf axils give rise to **vegetative branch** stems or to **flowers. Terminal buds** are present at the tips of the main stem and branches and contain the **apical meristem** tissues. The shoot originates in the embryo at the end opposite the root and develops a complex **shoot apex,** different from that of the root (see Table 6-1).

Table 6-1: The Shoot Apex of Roots and Stems

	Roots	*Shoots*
Apical meristem	covered by a cup-shaped **root cap**	exposed, no cellular cap
Apical appendages	absent	produce **primordia** (leaves and buds)
Orientation of cell division	occurs in all planes	oriented; two locations: inner (**corpus**) and an outer (**tunica**)
Zones	separate areas of division, elongation, and differentiation	no such recognizable zones present

The growing point of the shoot—the **apical meristem**—is surrounded by developing leaves (**leaf primordia**) that have in their axils **bud primordia.** The buds are of two kinds: Some are **vegetative** and will develop into leafy branches; others contain rudimentary **reproductive** tissues and will produce flowers.

Auxins (hormones) produced in the tip ordinarily inhibit the growth of the lateral bud primordia, and they remain dormant for some time; if the apical bud exerts **apical dominance** over the lateral buds, the plants produced will be conically shaped with a single leader and shorter lateral branches. If apical dominance is weak, axillary buds develop into branches soon after the terminal shoot elongates, resulting in a plant with many, branched stems, none the clear leader. Auxins produced in the leaf primordia control the elongation and differentiation of the primary meristems.

Primary Growth of Stems

Primary meristems

Stems, like roots, grow in length by division and elongation of cells at their tips. The youngest cells of stems (but not roots) are organized into two zones: the **tunica** and the **corpus.** In the tunica, cell divisions are perpendicular to the stem axis and give rise to a sheet of tissue several layers thick that covers the outside of the tip. Cell divisions in the corpus are in all directions and produce an interior mass of cells. Derivatives of cells in both tunica and corpus continue to divide and produce three recognizable **primary (transitional) meristems**—protoderm, ground meristem, and procambium—which, as they elongate and differentiate, create the three **primary tissue systems**—**dermal, ground (fundamental),** and **vascular.**

Cell divisions of the apical meristem give rise to leaf primordia close to the tip and so consistently, one after the other, that nodes and internodes can't be distinguished until elongation and differentiation start. At the base of the leaf primordia in the internodal region in monocot shoots, a zone of meristematic cells (an intercalary meristem) remains undifferentiated and retains the ability to divide throughout the life of the plant, causing elongation of the monocot leaves from the base upward.

Steles

The central cylinder of a primary plant body is called the **stele.** It consists of the primary xylem and phloem tissues together with any pith that may be present. Three types occur: 1.) **protostele,** the simplest, is a solid vascular core and is found in primitive vascular plants and the roots of eudicots (but not monocots); 2.) **siphonostele,** a hollow cylinder of vascular tissue surrounding a central core of pith is common in ferns; 3.) **eustele,** is a system of separate vascular bundles surrounding a pith and is the type in almost all seed plants.

Types of primary bodies

Herbaceous plants—in contrast to trees and shrubs—are composed essentially of primary tissues. If secondary growth does occur, the tissues are used for rigidity and not conduction. Two basic variations in the primary body of eudicots include a hollow cylinder of xylem, cambium, and phloem surrounding a central pith and, in others, a system of discrete vascular bundles, also with xylem, cambium, and phloem, arranged in a regular pattern between the pith and the cortex (see Figure 6-1). The primary body of monocots consists of vascular bundles, with no cambium, scattered in an undifferentiated parenchyma called ground tissue.

Secondary Growth of Stems

An aquatic plant is buoyed by the water in which it grows, and its structural needs are simple. Land plants, however, require a structural support system. During the course of evolution when plants developed the ability to synthesize lignin—the polysaccharide that gives rigidity to the cell walls of wood—large, erect bodies were achievable, and their possessors became highly successful in colonizing the land. In modern plants, lignified wood cells are the secondary xylem cells. Most of the primary tissues outside of the vascular cambium are destroyed by the sideways push of the new cells, and a new group of secondary tissues—the bark—replace them (see Figure 6-1).

Formation of the secondary plant body

During formation of the primary body, many plants retain meristematic tissues among differentiated ones. When stimulated to divide, these meristems, called cambia, produce new cells that, together with the remaining primary tissues, form the secondary (woody) plant body.

Vascular cambium

The vascular cambium lies between the primary xylem and phloem. It consists, accurately, of only one layer of cells, but the first cells it produces cannot be distinguished from cambial cells so the narrow area is sometimes referred to as the "cambium" or the "cambial zone."

Two kinds of meristematic cells, called initials, are recognizable in the cambium: fusiform and ray initials. The fusiform initials are elongated vertically in the stem and have tapering ends. They divide to produce the conducting cells of both the xylem and the phloem (xylem toward the inside of the stem, phloem toward the outside).

Considerably more xylem cells than phloem cells always are produced. The ray initials are smaller, more cuboidal and produce parenchyma in rows radiating out from the center of the stem. The bands of parenchyma, called rays (vascular rays), conduct water and dissolved materials laterally in the stem.

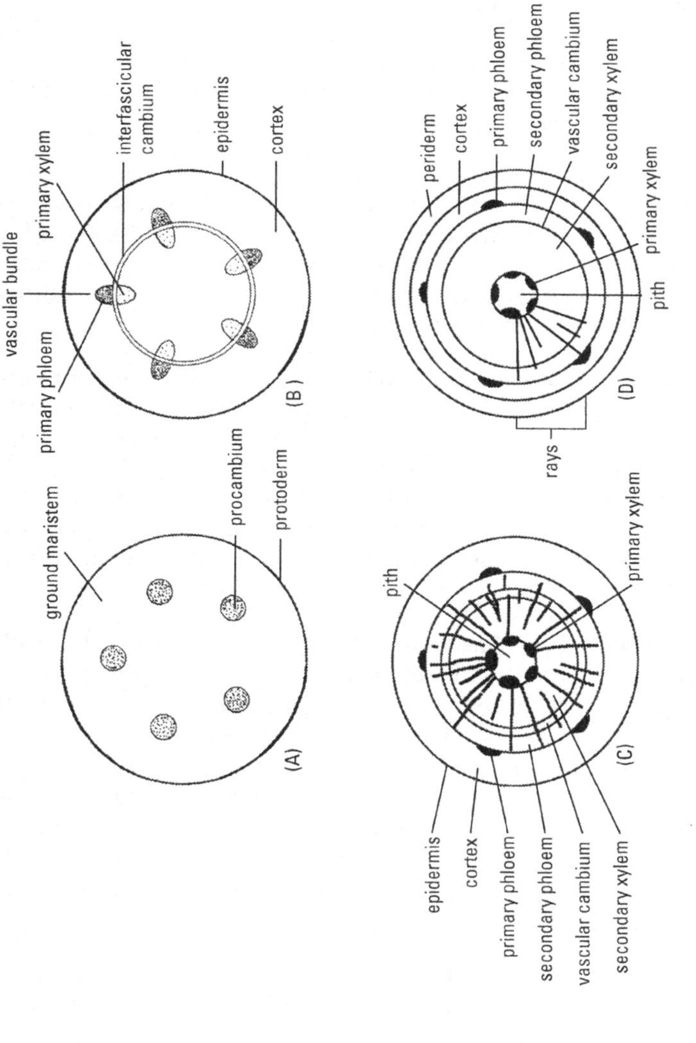
ground maristem
procambium
protoderm
(A)
primary phloem
vascular bundle
primary xylem
interfascicular cambium
epidermis
cortex
(B)
pith
primary xylem
epidermis
cortex
primary phloem
secondary phloem
vascular cambium
secondary xylem
(C)
rays
periderm
cortex
primary phloem
secondary phloem
vascular cambium
secondary xylem
primary xylem
pith
(D)

Figure 6-1

Wood: Secondary xylem

The structure of wood varies from species to species and between major groups. A common categorization separates the *softwoods* of gymnosperms from the *hardwoods* produced by angiosperms. (These are not very good descriptive terms because of the great variability in density among species in both groups, but the groups do differ in the kinds of cells in their wood.)

Gymnosperm wood. Softwood lacks vessels and is composed almost entirely of tracheids. The rays are ribbon-like structures of parenchyma, one cell wide and only a few cells deep. Vertical **resin ducts** or canals are characteristic of gymnosperms. The ducts intercellular spaces lined with parenchyma tissue, the cells of which secrete resin into the cavity in response to wounding.

Angiosperm wood. Hardwoods are harder than most softwoods because of the numerous fibers present. The usual conducting cells (tracheids and vessel segments), scattered parenchyma, and ray parenchyma are present in the wood. Some dicot (eudicot) species have resin and resin ducts, but other substances—latex (rubber), for example—are more commonly secreted in angiosperms in response to wounding. The rays of the hardwoods usually are multiseriate (many cells in width) and hundreds of cells deep.

Growth rings. In climates that alternate favorable with unfavorable seasons for plant growth, the xylem cells produced by the cambium vary in size throughout the growing season, resulting in rings with discernible differences. If there is one growing season per year, the rings are *annual rings* and a simple count gives the age of the tree. Other circumstances may cause the cambium to stop and start growth—forest fires, volcanic eruptions, defoliating caterpillar outbreaks, or extreme drought, for example—with *false annual rings* the result. The science of **dendrochronology** is the study of growth rings to date past events and climates.

Characteristics of wood. Many descriptive names are commonly applied to easily seen features of wood. Wood in the center of trees is called **heartwood** and it often is discolored by accumulations of tannins, gums, and oils carried there and stored in balloon-like outgrowths called **tyloses** that fill and plug the vessels. **Sapwood** is the youngest, last-formed xylem. All of the heartwood can decay, leaving a hollow trunk, and the tree will remain alive and healthy if the sapwood is intact.

Bark: Secondary phloem and periderm

All tissues from the vascular cambium outwards collectively are the **bark** of a woody plant. Almost all are secondary in origin and are produced after divisions in the vascular cambium start.

Phloem. Secondary phloem cells are produced by the vascular cambium at the same time as secondary xylem cells, but in fewer numbers. Their outward growth pushes the primary phloem cells against the cortex, breaking most and leaving only the thicker-walled fibers as remnants. Ray parenchyma cells initiated by the cambium give rise to **phloem rays** and, towards the center of the stem, **xylem rays.** The rays are the primary avenues of lateral movement of materials from the vascular vertical conduits that lie close to the cambium.

Periderm. Changes occurring in the cortex and the epidermis replace the protective layers of primary tissues with **periderm,** which consists of three tissues: the **cork cambium (phellogen), cork (phellem),** and the **phelloderm.** Both cork and phellloderm arise in the cork cambium, but differ structurally and functionally. Cork, with heavily suberized, tightly packed, dead-at-maturity cells, lies to the outside of the cork cambium and serves to waterproof, insulate, and protect the underlying stem tissues. Phelloderm, in rows to the inside of the cork cambium, remains alive and carries on routine metabolic functions (including photosynthesis in some green-barked trees). **Lenticels,** small openings in the bark, permit diffusion of gases. They are raised areas filled with loose parenchyma cells that are in direct contact with the atmosphere on one side and with cortical tissues on the other.

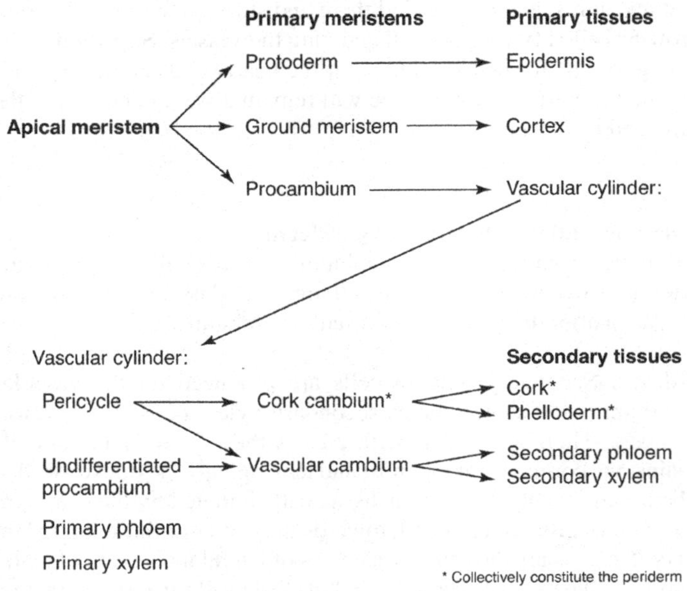

Figure 6-2

Specialized Stems of Angiosperms

Not all stems conform to the definition given at the beginning of this chapter: Many stems are not aerial, some grow flat on or in the ground, some are leafless, others look like leaves. The more common variations (and their names) are included in the discussion of leaves in the following chapter.

CHAPTER 7
THE SHOOT SYSTEM: LEAVES

Leaves Are Specialized Organs

Over the geological eons, as structural changes in stems and roots made larger and taller plants possible, the branching systems that served as the first leaves—the sites of photosynthesis—changed, too. Effective mechanical tissues, especially those that function with dead cells, make poor tissues in which to conduct the bio-chemical process of photosynthesis; hence, the success of parenchyma-filled leaves.

Leaves arose independently in the lineages of flowering plants and other major plant groups and, in most instances, became the specially modified sites for photosynthesis. Modern leaves come in all sizes and shapes, and their diversity provides an oft-used means to identify kinds of plants. While genetics determines the way leaves are arranged on stems (their **phyllotaxy**) and their basic form, physical factors of the environment such as light and water are responsible for many of the variations in their appearance.

External Features of a Leaf

Figure 6-1 in the preceding chapter illustrates a few basic terms that are applied to leaves. Taxonomists use an inordinate number more terms as a means to separate and name plants. The terminology applied to the way leaves are attached to the stem, for example, includes **alternate**—the arrangement shown in Figure 7-1—as well as **opposite** and **whorled** and is based on the number of leaves attached at each node: one (alternate), two (opposite), and three or more (whorled). If a single blade is attached to a petiole, as in Figure 7-1, the leaf is **simple;** if the blade is divided into two or more individual parts, the leaf is **compound** and may be **pinnately** or

palmately so depending upon how the **leaflets** (the individual separate units of the blade) are attached to the extension of the petiole (the **rachis**). Other standard terms are used for venation, overall shape, shape of the tip, condition of the edge of the blade (toothed, smooth, lobed), hairy (what kind of hairs) or smooth (on both upper and lower surfaces or just on one) and more.

Origin of Leaves at the Shoot Apex

Leaves arise in the shoot apex of stems in cells immediately below the protoderm. Division and expansion of the cells in this area result in a **leaf primordium** in which meristematic regions soon become identifiable in the upper and lower regions of the tissue destined to become the blade. A strand of procambium from the shoot, the **leaf trace,** makes connection with differentiating vascular tissues of the primordium thus assuring the continuity of the conducting tissues throughout the plant. The area on the vascular cylinder of the stem where the leaf trace diverges into the leaf primordium is called a **leaf gap,** a confusing name; it is not a hole but an area filled with parenchyma cells. "Gap" refers to the absence of xylem and phloem cells at this point in the vascular cylinder.

The tissues of the evolving blade develop faster on the lower (**abaxial surface**) than those on the upper (**adaxial surface**) with the result that the primordium bends inward towards the shoot apex. The elongating primordia arch over and protect the apical meristem of the shoot. Cells divide and elongate in the primordium, differentiating downward from the tip and the intercellular spaces characteristic of the mature leaf soon appear among the young blade tissues. Cell divisions cease when the leaf is less than full size, and subsequent enlargement consists of elongation and expansion of cells and intercellular spaces. Leaves thus have *determinate* growth, whereas the apical meristem, with its cells that continue to divide indefinitely, has *indeterminate* growth.

Internal Structure

The standard leaf has three tissue regions: the epidermis, the mesophyll, and the vascular bundles or veins (Figure 7-1).

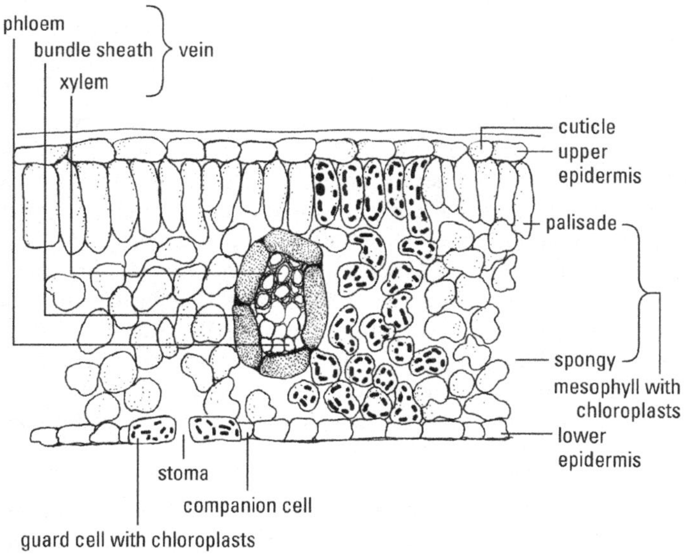

Figure 7-1

Epidermis
The epidermis of leaves is a continuous layer of cells on all surfaces of the leaf, unbroken except for pores, the **stomata** (**stoma,** singular), which facilitate the exchange of gases between the interior of the leaf and the atmosphere. The parenchyma cells of the epidermis fit together like paving stones and generally contain no chloroplasts

except for those in the guard cells of the stomata. A **cuticle** composed of **cutin** and wax is deposited on the outer primary walls of the epidermal cells. It varies in thickness among different kinds of plants. Hairs or scales—called **trichomes**—are extensions of epidermal cells and are present on many leaves. **Glands** associated with trichomes often produce substances repugnant or toxic to herbivores. The physical presence of a tangle of trichomes on the surface of a leaf also deters many animals from eating or using the leaf.

Stomata consist of two kidney-shaped **guard cells** surrounding an opening, the **stoma,** and usually two to four **subsidiary cells**—ordinary parenchyma cells shaped to fit around the guard cells so no holes are left in the epidermal covering. (Note that "stoma" refers both to the small pore alone as well as to the entire apparatus of guard cells plus the pore.) The walls of the guard cells facing the stoma are thicker than the opposite walls and more elastic. When the guard cells fill with water (become turgid) the thinner walls elongate faster than those facing the pore, thus pulling the latter walls away from one another and opening the pore. Conversely, when the cells lose water and contract (become flaccid), the walls relax and the pore closes. The stomata regulate the passage of most of the water from the leaves and the movements of air in and out.

Depending upon where the plant lives and how its leaves are oriented, stomata may be present on both the upper and lower leaf surfaces, on one or the other exclusively, or be lacking from the leaves entirely, the latter case being characteristic of submerged aquatic plants.

Mesophyll

The mesophyll tissue forms the bulk of most leaves and the chloroplasts in its cells are the principal sites of photosynthesis. The mesophyll is sandwiched between the epidermal layers. In leaves held horizontally on stems and in which there is a discernable top and bottom, the upper and lower mesophyll cells have different shapes whereas in leaves held vertically, the mesophyll is uniformly the same throughout.

If the mesophyll is differentiated, the upper layer is called the **palisade mesophyll** and consists of closely packed columnar cells with their long axis at right angles to the leaf surface. The lower tissue, called **spongy mesophyll,** is made of irregularly shaped cells, loosely arranged with much intercellular space. While both mesophyll types contain chloroplasts, the palisade has more than does the spongy mesophyll. The mesophyll, therefore, is a type of chlorenchyma—chloroplast-containing parenchyma. The spongy mesophyll with its air spaces is, additionally, an aerenchyma.

The wet surfaces of the mesophyll cells are the sites of water loss and gas exchange; the stomata are merely the gates through which the water and gases pass to the outside.

The mesophyll contains strengthening tissues, primarily around the veins, but also in scattered batches throughout the mesophyll. Sclereids are especially common and almost always collenchyma cells are used to strengthen veins. Fibers are common in the leaves of monocots.

Veins (vascular tissue)

Veins penetrate all parts of the leaf, forming a network that connects the leaf through the petiole to the vasculature of the stem and thereby to the root as well. **Primary xylem** cells occupy the upper part of the vein and **phloem** cells the lower. The vascular tissues are surrounded by a **bundle sheath** one or two layers thick, composed of fibers in the smaller veins and parenchyma in the larger.

Fibers and collenchyma are present in and around the veins and give strength to them and to the leaf as a whole. Bundle sheath extensions connect the bundle sheaths to either or both epidermises giving added stability to the blade. The large veins branch repeatedly becoming smaller each time they divide until they ultimately end with only one or two tracheids at the vein ending. Here the mesophyll cells are in direct contact with—or at most one or two cells away from—the raw materials carried in the xylem and used for photosynthesis. The phloem is equally convenient for export of photosynthetates. The bundle sheaths insulate the conducting cells and ensure the retention of materials in the pipeline.

The veins of tropical grasses and other plants with C4 photosynthesis are surrounded by two cylinders, the inner of thick-walled bundle sheath cells, the outer of thin-walled mesophyll cells. C4 plants are said to have a **Kranz** (from the German word for wreath) anatomy because of these. In addition, no distinct palisade or spongy mesophyll zones are present in the C4 leaves.

Leaf Abscission

All leaves have a definite life span and are dropped following receipt of internal or environmental signals. The process is termed **abscission** and is facilitated by the formation of an **abscission zone** at the base of the petiole. Plants that drop all of their leaves within a short time resulting in a temporarily bare, leafless plant are called **deciduous** plants. Those that drop leaves a few at a time throughout the life of the plant are called **evergreen** plants (they appear to be fully leafed at all times).

Hormones trigger the formation of the abscission layer. Severance of the leaf is aided by anatomical changes in the abscission zone where two tissue zones differentiate; the one nearest the stem accumulates suberin in the cell walls—blocking the flow of materials—while cells of the separation layer on the blade side simply disintegrate. The suberized zone left on the stem after the leaf falls is called the **leaf scar;** visible within it are **bundle scars,** the remnants of the vascular strands.

Leaf Movements

Some leaves have anatomical specializations that make possible responses almost as fast as those of animal movements. In one sensitive plant, a *Mimosa*, touching the leaves causes a change in the permeability of the membranes of the large, thin-walled parenchyma cells in the **pulvini** (singular, pulvinus)—the swollen glands at the bases of the petioles—and an almost instantaneous water loss. The whole leaf droops as the pulvini cells become flaccid. Pulvinus-mediated movements of other taxa are slower, but also serve to move the petioles.

Some plants change the position of their leaves daily, dropping them to a vertical position at night, elevating them back to horizontal at dawn. These sleep movements, one of many circadian rhythms in plants, are called **nyctinastic** and the process, **nyctinasty.**

Insectivorous plants such as the Venus flytrap have perfected a combination of anatomical and physiological leaf specializations to attract, catch, and digest insects.

Leaves and the Environment

Light effects

Sun and shade leaves. The architecture of leaves changes depending upon the light intensity in which they grow, even on the same plant. Sun leaves usually are smaller and thicker with more and better defined palisade cells, and more chloroplasts. They frequently have more hairs as well. Sun leaves rarely have chloroplasts in their epidermal cells, but chloroplasts are common in the epidermises of shade leaves.

Day length. The presence or absence of light (as well as particular wavelengths) influences the production of plant hormones and the development of plant organs. For example, leaves do not develop normally in the dark, and chloroplasts don't turn green unless exposed to light; the tissues are yellow and said to be **etiolated.** The duration of the light also influences the shape of many kinds of leaves; leaves produced during the short days of spring are different from those produced during the long days of summer.

Water effects

The presence or absence of water in the environment has profound effects on the structure of plant stems, roots, and leaves, so much so that three types of plants are recognized based on the water content of the soil: xerophytes, mesophytes, and hydrophytes. In this categorization, the common plants, mesophytes, live where water is neither abundant nor limited, i.e. in a mesic (meaning middle) environment. Xerophytes (xeric meaning dry) are adapted for life in arid regions while hydrophytes (hydric meaning water) live in water or else have their roots in wet soil. While it is tempting to attribute structural adaptation to one factor of the environment alone, in truth, all of the physical and biological factors of the environment undoubtedly contribute. Because of the role of water in plant metabolism, however, water and its availability clearly control the structures as well as the functioning of plants.

Specialized Leaves and Stems

Although typical shoots are erect with photosynthetic leaves, over evolutionary time a great assortment of modifications of the basic body plan have arisen. Some clearly benefit storage of materials, others assist in **vegetative reproduction** (reproduction without seeds), various alterations deter herbivores, and many are simply innovations in ways to hold the shoot upright. The most bizarre of all may be the leaves of the **insectivorous** plants that are modified to ensnare and digest hapless insects and other small organisms. Some drown their victims in vase-like rainwater-filled petioles while others glue them to the leaf with sticky digestive enzymes. The Venus' flytrap, on the other hand, snaps its leaves together rapidly enough to enclose the unlucky insect that alights on the trigger hair.

You can see many of the modifications in common garden and edible plants. For example:

- **Bulbs** are underground buds with the stem reduced to a small knob on which fleshy storage leaves are clustered (e.g. dry onions).
- **Tubers** are fleshy underground stems modified to store starch (e.g. white, or Irish, potatoes). The "eyes" are the nodes with an axillary bud in each (the peel is periderm tissue). Sweet potatoes are *roots*.
- **Hizomes** are horizontal underground stems with nodes, internodes, dry scale leaves, and adventitious roots (e.g. fresh ginger "roots" sold in grocery stores are rhizomes). Canna lilies, iris, and many grasses have rhizomes with which they are propagated.
- **Corms** are upright underground fleshy stems covered by leaves reduced to dry, covering scales (e.g. gladiolus and crocus). Note that corms store reserve food in stem tissue, and bulbs in leaf tissue.
- **Thorns** are woody, sharply pointed branch stems (e.g. honey locust).
- **Spines** are small, unbranched, sharp outgrowths of leaf tissue in which the parenchyma is replaced by sclerenchyma (e.g. cactus).
- **Prickles** are small pointed outgrowths from the epidermis or cortex of the stem (e.g. rose and raspberry).

- **Cladophylls** are flattened main stems that resemble leaves (e.g. butcher's-broom, greenbrier, and some orchids). Edible asparagus shoots left to grow produce many small fern-like cladophylls.
- **Stipules** are paired scales, glands, or leaf-like structures at the base of the petiole formed from leaf or stem tissue (e.g. black locust).
- **Bracts** are modified leaves at the base of flowers or flower stalks. Some are highly-colored and resemble petals (e.g. the red "petals" of poinsettia are bracts surrounding the small, yellow flowers).
- **Tendrils** can be exclusively leaf tissue (e.g. pea leaflets, nasturtium petioles, or cucumber leaves that twine and aid in supporting the shoots) or they can be modified special shoots with thin, modified stems (e.g. morning glories, grapes, and Boston ivy).
- **Stolons**, sometimes called *runners*, are thin, above-ground, horizontal stems of indeterminate growth and long internodes that grow out from a parent plant and produce young plants at their tips (e.g. strawberry plants, and a host of the most pernicious garden weeds).

CHAPTER 8
THE SHOOT SYSTEM: FLOWERS, FRUITS, AND SEEDS

Features of Flowers

Flowers arise from apical meristems similar to vegetative shoots but, unlike them, have determinate growth. The floral primordia develop into four different kinds of specialized leaves that are borne in whorls at the tip of the stem (see Figure 8-1). The two outer whorls are sterile, the inner two fertile. The first formed outer whorl—the **calyx**—is the most leaflike and its individual parts, the **sepals,** often are green. The **petals** of the next whorl, the **corolla,** frequently are brightly colored and in a majority of flowers retain some semblance to leaves. (Together the calyx and the corolla are called the **perianth.**) The next two whorls, the **androecium** and the **gynoecium,** are composed of highly modified reproductive structures that have lost their leaf-like appearance. The androecium is composed of **stamens** and the gynoecium of **carpels.** (**Pistil** is sometimes used as the term for a single carpel or a group of fused carpels.) The stamens are microsporophylls and have a stalk, the **filament,** at the top of which the **pollen**-bearing **anthers** are located. A carpel is a megasporophyll and has as its base an enlarged **ovary** from which the **style** bearing a **stigma** arises. The whorls are attached to the **receptacle** area at the end of the flower stalk or **pedicel.** Some flowers arise singly, but more are produced and arranged in groups called **inflorescences.** The stalk of an inflorescence is the **peduncle** and the extension of the axis in the inflorescence is the **rachis,** to which the pedicels of the individual flowers are attached.

It is tempting, but botanically incorrect, to refer to the stamens as "male" floral parts and the carpels as "female" since both are part of the *sporophyte* generation and only gamete-producing plants, i.e. gametophytes, have gender. The male gametophytes of angiosperms develop within the anthers, the female in the ovules.

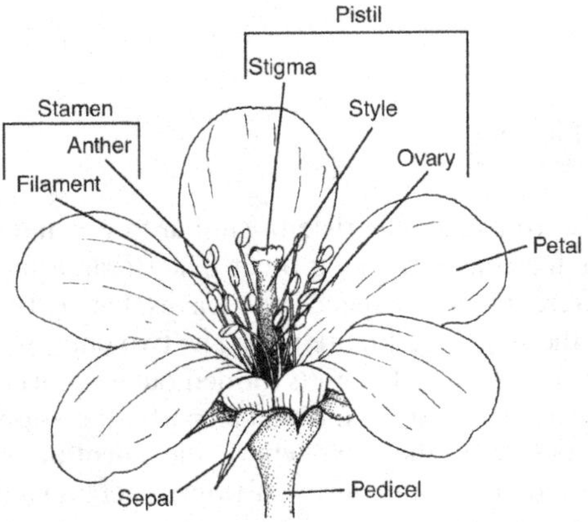

Figure 8-1

Seed and Fruits

Seeds develop from ovules in the ovary, and at maturity consist of an **embryo** and a reserve food supply surrounded by a protective covering, the **seed coat.** The diversity of flowering plants assures diversity among their seeds, but, unlike fruits, which have numerous variations, structural plans for seeds are few. The reserve food can be stored either in or out of the embryo and the **cotyledon(s)**—the seed leaves—can remain either below ground or be elevated above the surface when germination occurs.

Fruits are ripened ovaries containing seeds with sometimes additional flower or inflorescence tissues associated with them. Only angiosperms produce flowers and fruits. From a botanical viewpoint, many of the foods we eat as vegetables are fruits, e.g. tomatoes, green beans, squash, eggplant, and peppers. Fruits apparently arose as a means not only of protecting the seeds, but as a way to ensure their dispersal.

Features of the Angiosperm Life Cycle

Like other plants, the angiosperms alternate a sporophytic generation with a gametophytic one, a **sporic meiosis** (see Figure 8-2). Angiosperm **sporophytes** are the common plants around us—trees, grasses, and garden vegetables. They produce through **meiosis** (reduction division) two kinds of spores in specialized structures of their flowers, microspores in the anthers and megaspores in the ovules contained within the ovaries. The gametophytes, which develop from the spores, are much reduced in size. The microgametophyte consists of only three cells and is the germinated pollen grain. The megagametophyte is the mature embryo sac, a seven-celled structure in the ovule surrounded by, and dependent upon, sporophyte tissue.

Development of the gametophytes

Haploid **microspores** develop from **microsporocytes** in the anthers and give rise to pollen grains containing two cells: the **tube cell** and the **generative cell.** At about the time of pollination, the latter cell divides and produces two **sperm.** This three-celled pollen grain is the immature male gametophyte (microgametophyte).

The female gametophyte, the **megagametophyte,** develops in the ovary at the same time the male gametophyte is developing in the anthers. While the process is exceedingly variable among taxa, about three-quarters of the flowering plants go through the following steps.

One **megasporocyte** is contained in each of the young ovules within the ovaries in the flower buds. The ovule is attached by a stalk, the **funiculus**, to the **placenta** on the ovary wall and, at this stage, is essentially a lump of tissue, the **nucellus,** covered by two tissue layers, the **integuments** (which wrap almost completely over the ovule but leave a small opening, the **micropyle,** at one end). Three of the haploid megaspores produced by the megasporocyte disintegrate almost immediately and the remaining one divides by three successive mitotic divisions to produce eight nuclei in an **embryo sac** within the elongated, swollen megaspore. The nuclei cluster, first in groups of

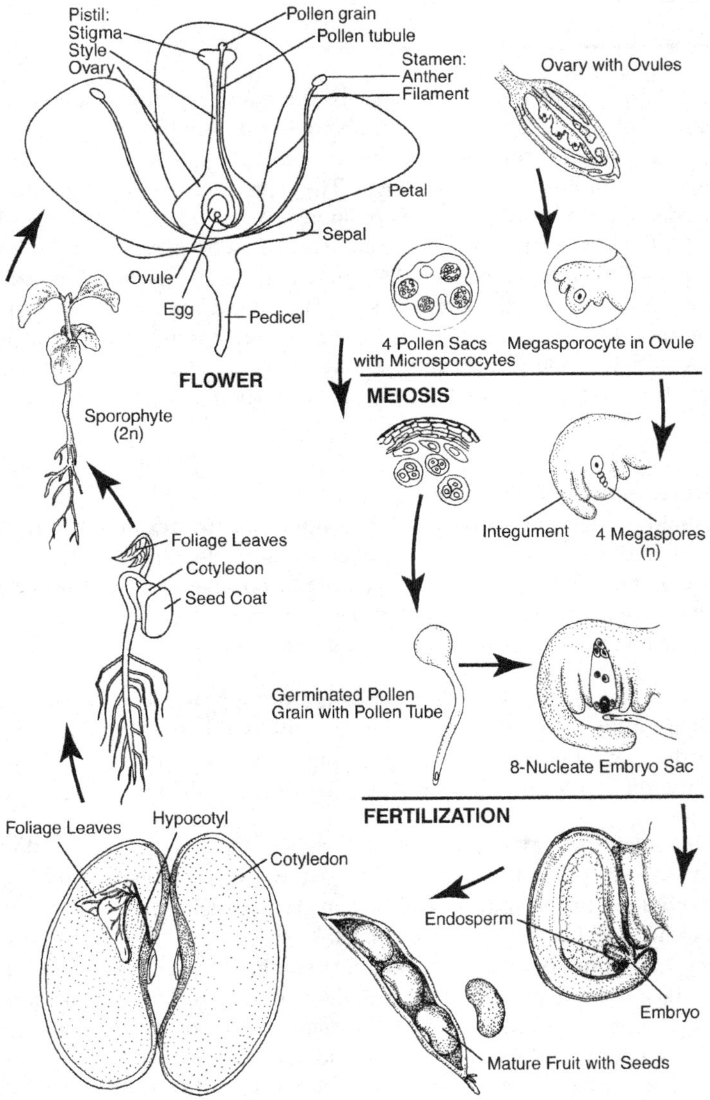
Pistil:
Stigma
Style
Ovary
Pollen grain
Pollen tubule
Stamen:
Anther
Filament
Ovary with Ovules
Petal
Sepal
Ovule
Egg
Pedicel
4 Pollen Sacs with Microsporocytes
Megasporocyte in Ovule
FLOWER
MEIOSIS
Sporophyte (2n)
Integument
4 Megaspores (n)
Foliage Leaves
Cotyledon
Seed Coat
Germinated Pollen Grain with Pollen Tube
8-Nucleate Embryo Sac
FERTILIZATION
Foliage Leaves
Hypocotyl
Cotyledon
Endosperm
Embryo
Mature Fruit with Seeds

Figure 8-2

four at either end of the sac and then one nucleus from each end, migrates to the center. The two migrating nuclei are called the **polar nuclei** and they form a **polar cell** when walls develop around them. Cell walls also form around the three nuclei left at the end of the cell opposite the micropyle, the **chalazal end.** The chalazal end cells are the **antipodals.** At the micropylar end, the three nuclei are organized into the **egg apparatus** and walls form around each of them also. One cell is the **egg** cell, the two others are **synergids** (helpers). All three look alike, but only the egg continues to develop; the synergids deteriorate as do the antipodals on the opposite end of the sac. The **embryo sac** at this stage is the female gametophyte (megagametophyte). Before further development can occur and seeds are produced in the ovary, two events must occur: pollination followed by fertilization.

Pollination

Pollination is the mechanical transfer of pollen grains from an anther to a **stigma,** the receptive end of a carpel. Pollination is accomplished by a variety of physical dispersal agents such as wind, water, and gravity or many kinds of animals including insects, bats, birds, and small rodents. The variations in floral structure are, in large part, adaptations to achieve pollination success. Most pollination is between flowers located on separate plants (**cross-pollination**), but in some taxa **self-pollination** occurs when pollen from the anthers falls on stigmas of the same plant.

Fertilization

If the pollen grain lands on the stigma of a genetically compatible flower, it absorbs moisture and a **pollen tube** emerges through a pore in the wall. The germinated pollen grain with its pollen tube and three nuclei is the **mature male gametophyte.** The tube grows downward toward the ovary through special tissues in the style, penetrates the embryo sac, usually through the micropyle (destroying a synergid in the process), and discharges its contents. The tube nucleus disintegrates while one of the sperm nuclei fuses with the egg nucleus, forming a **zygote.** The other sperm fuses with the polar cell, forming the **endosperm nucleus.** In other words, a **double fertilization** occurs: Both sperms fuse with embryo sac nuclei. Double fertilization is a characteristic of the angiosperms and results in a **polyploid** endosperm tissue. (Polyploidy refers to the number of sets of chromosomes the cell contains; plants with more than the diploid two sets are polyploids. The endosperm tissue may be *triploid* [3*n*] or more depending upon the species.)

If no pollen tube and its contents reach an ovule in the ovary, the ovule aborts with no further development. Lacking chemical signals (hormones) from a developing seed, the ovary, too, may wither and die. If double fertilization does occur, the ovule develops into a **seed** and the entire ovary into a **fruit.**

Seed Structure and Development

Following the fertilizations in the embryo sac, the zygote divides repeatedly by mitosis and differentiates into an **embryo.** The endosperm nucleus also divides by mitosis and forms the **endosperm** tissue, which provides food for the developing embryo.

The early embryo is linear with apical meristems on either end and one or two seed leaves or **cotyledons.** The axis below the cotyledons is called the **hypocotyl,** at the tip of which is the **radicle** that gives rise to the primary root of the seedling. The axis above the attachment of the cotyledons is the **epicotyl,** which also ends in an apical meristem. In some seeds, the first foliage leaves are formed in the seed. The area above the cotyledons is thus a miniature shoot and is called the **plumule.** In some taxa, food for the embryo remains within endosperm tissue and the cotyledons serve as organs of absorption. In others, food moves directly into the embryo and is stored within the cotyledons, leaving only a minuscule endosperm. In still another variation, the nucellus (the megasporangium wall) enlarges and becomes a storage tissue called **perisperm.** The integuments harden into the

seed coat as the embryo matures. The scar left on the seed coat by the separation of the funiculus from the integuments is called the **hilum.** Often the micropyle remains visible near the hilum.

Angiosperms traditionally have been separated into two major categories on the basis of the number of cotyledons they possess: **monocots** (mono = one; cotyledons = seed leaves) and **dicots** (di = two; cotyledons = seed leaves). Monocots are the grasses, sedges, lilies, and their relatives, while **dicots** constitute the rest of the flowering plants. This artificial separation currently is being replaced by a more natural classification in which the *monocots* are retained as a natural group, but the dicots are separated into the **eudicots** (eu = true, dicots) and the **magnoliids.** The latter is a small group of very primitive angiosperms ancestral to both monocots and eudicots and has both woody representatives—**woody magnoliids** (trees like *Magnolias*, tulip trees, and laurels)—and **paleoherbs** (herbaceous plants such as members of the pepper and water lily families). The eudicots constitute about 97 percent of the angiosperms, while the magnoliids make up the other 3 percent.

In monocot seeds the single cotyledon usually digests and absorbs food from the endosperm and translocates it to the embryo. In grasses, the cotyledon is called the **scutellum.** In grasses, also, there is a protective sheath called the **coleoptile** over the plumule and another, the **coleorhiza,** surrounding the radicle.

Seed germination

Internal signals shut off growth of the embryo at a certain size and the seed goes into a period of **dormancy.** In some plants, dormancy lasts only as long as it takes the seed to be dispersed from the ovary. In others, dormancy may last for long periods until either external or internal signals (or a combination of both) initiate further growth. Environmental factors of chief importance to initiate growth are water, light, and temperature.

Seedling growth

When an embryo resumes growth, stored food provides the energy for seedling development—the roots first, followed by elongation of the photosynthetic shoots.

Eudicot development. Aboveground growth in eudicots takes one of two general patterns: **epigeous** or **hypogeous.** In epigeous growth, the hypocotyl elongates, pulling the plumule and cotyledons above ground; in hypogeous growth, the cotyledons remain below ground because the epicotyl grows faster than the hypocotyl and pulls the plumule erect.

Monocot development. Monocots develop with two different general patterns: one for the grasses, one for the rest of the group. In most of the monocots (but not grasses), after the radicle has pushed out of the seed coat, the first shoot structure to emerge is the cotyledon, which arches upward with the remainder of the endosperm and the seed coat still attached. It elongates above ground and is photosynthetic until the true leaves develop.

In the grasses, the sheaths around both the shoot and the root tip must be penetrated by the roots and the shoots. The root sheath, the coleorhiza, grows faster than the radicle for a short time, but when it stops growing, the radicle emerges and forms an anchoring primary root. The shoot sheath, the coleoptile, moves upward to the soil surface through elongation of the first internode of the stem (called the *mesocotyl*), and when it reaches the surface it stops growing. The plumule then pushes through into the air. At about the same time, buds of adventitious roots begin to grow and, by the time the seedling is erect with a few true leaves, it already has adventitious roots growing downward from its first node. The primary root system is short lived and dies soon after its establishment and the adventitious root system becomes the principal absorbing and anchoring system for the new grass plant.

CHAPTER 9
ENERGY AND PLANT METABOLISM

Energy Defined

Energy is the capacity to do work. It is the flow of energy that makes the work of life (metabolism) possible. From the smallest unit of life—the cell—to the largest organism, all living things obtain, modify, expend, and release energy and in doing so adhere to very explicit energy laws of the universe.

Of the tremendous amount of energy received from the sun, less than 1 percent is captured by green plants and converted through photosynthesis to usable energy. Some solar energy drives the geological processes that make the earth habitable for organisms then is lost back into space. Fully a third never makes it through the Earth's atmosphere to reach the surface. The small percentage that is captured in photosynthesis runs the entire Earth ecosystem.

The Laws of Thermodynamics

Universal laws of energy exchange, the **laws of thermodynamics,** govern all interactions among organisms (and all matter). Two are especially important in explaining how organisms manage their energy needs.

- **First Law of Thermodynamics**—the conservation of energy—simply states that while the form of energy can be changed, energy itself can neither be created nor destroyed. Energy exists in two forms: **potential energy** (stored energy *available* to do work) and **kinetic energy** (the energy used to *do* work).
- **Second Law of Thermodynamics**—the law of **entropy** (disorder)—in brief, states that chemical reactions run downhill,

i.e. the products of the reaction always have less potential energy than the original reactants. Entropy measures the randomness or disorder that forever increases in systems to which no energy is added.

In each energy exchange—from the first photosynthetic reaction to the last in the food web where carnivores dine on one another—energy escapes, primarily as heat. Although the heat energy remains in the system (fulfilling the First Law of conservation), the energy is no longer available to do work, hence it is "lost" to further metabolism. Each of the exchanges is **exergonic** (ex = out; energy out). The heat of the system rises in proportion to the loss of potential energy. To maintain the organized systems like organisms, therefore, energy is added constantly in a series of **endergonic** (end = in; energy in) processes.

The Chemistry of Energy Use

Metabolism

The total of all the chemical reactions occurring in living organisms is **metabolism** and the sequences in which they occur are the **metabolic pathways.** Cells can metabolize because they are isolated systems, separated from their environment by **membranes.** Organisms, and the cells of which they are composed, use enzymes to regulate the reactions and utilize energy carriers to move energy among the parts of the system. Metabolic reactions are linked so that exergonic reactions supply energy for the endergonic.

Oxidation–reduction reactions

Most of the energy exchanges in plants are chemical reactions that involve the exchange of energy between one set of chemical bonds and another. These, for the most part, are **oxidation-reduction** reactions (commonly called **redox** reactions). In oxidation, electrons are *lost* from an atom or molecule, which is then said to be **oxidized.** (The term "oxidation" is used because oxygen is often the electron acceptor.) Reduction always accompanies oxidation and is the acceptance or *gain* of electrons by another atom, which, thereby, is **reduced.** As electrons are lost, so is energy, thus oxidized molecules have *less* energy than reduced molecules that *gain* energy as they receive electrons.

In organisms, electrons rarely move alone and are usually accompanied by a proton (a hydrogen atom and its single electron). Oxidation, accordingly, entails the removal of a hydrogen atom, and reduction entails the addition of hydrogen atoms.

The collections of biological molecules within the cells are *reduced* and electron-rich and so have relatively weak chemical bonds. In the surrounding environment, most of the molecules are *oxidized* (and electron-poor) and have much stronger bonds. Living organisms thus continually add energy to their systems, to prevent operation of the Second Law of Thermodynamics, i.e. to avert becoming a disordered collection of oxidized molecules.

Photosynthesis and respiration are both oxidation-reduction processes. Photosynthesis requires the input of energy, while respiration releases energy; photosynthesis is thus an *endergonic* process, *respiration* an exergonic.

Energy Regulators: Enzymes and ATP

Enzymes

If all the energy in a reaction were released at the same time, most of it would be lost as heat—burning up the cells—and little could be captured to do metabolic (or any other kind) of work. Organisms have evolved a multitude of materials and mechanisms—such as enzymes—that control and permit the stepwise use of the released energy.

Enzymes control the state of energy a molecule must attain before it can release energy and are the chief **catalysts** of biochemical reactions. They are neither consumed nor changed in the reactions. Basically, enzymes reduce the **activation energy** needed to start a reaction by temporarily bonding with the reacting molecules and, in so doing, weakening the chemical bonds.

Almost all of the over 2,000 known enzymes are proteins, nearly all of which operate with **cofactors**—metal ions or organic molecules (**coenzymes**). The enzymes act in series with each enzyme catalyzing only a part of the total reaction (which is why there are so many of both enzymes and cofactors). If the same type of reaction occurs in two different processes, each requiring the same enzyme, two different but structurally similar enzymes are used. These are called **isozymes,** and each is specific for its own process.

Two different structural models are used to explain why enzymes work so efficiently. According to the **lock-and-key model,** there is a place in the enzyme molecule, the **active site** (the lock), into which the **substrate** (the key) fits by virtue of the latter's electrical charge, size, and shape. In actuality, however, the connection appears to be much more flexible than this model permits. The **induced-fit model** takes this into account and states that although size and shapes are comparable, the active site is flexible and appears to adjust to fit the substrate. In so doing, it tightens the connection when the molecules come together and initiates the enzymatic reaction. However it works physically, chemically the enzyme-substrate relationship is exact and specific, one enzyme for each substrate.

ATP (adenosine triphosphate)

Energy is the currency of the living world and ATP, like the coins that change hands in our economy, is the means through which energy is circulated in and among cells; it is the most common **energy carrier.** ATP is a nucleotide composed of adenine, the sugar ribose, and three phosphate groups. Its value as an energy carrier lies in the two easily broken bonds that attach the three phosphate groups to the rest of the molecule. These bonds are inappropriately called **high energy bonds;** they have ordinary energy values, but are weak and so easily split. Hydrolysis of the molecule (catalyzed by ATPase) breaks the terminal weak bond releasing energy, an inorganic phosphate (P_i) and ADP (adenosine diphosphate). Sometimes the reaction repeats, and the second bond also is broken releasing more energy, another P_i and ADM (adenosine monophosphate). The ADP is recharged back to ATP in cellular respiration. ATP is also made during photosynthesis.

ATP is indispensable for short-term energy use, but not useful either for long-term energy storage or for processes requiring large amounts of energy. The former needs are met in plants chiefly by starch and lipids, the latter by sucrose.

Movement of Materials in Cells

Cells are bathed in a watery matrix and conduct most of their reactions in a similar watery fluid—a **solution** in which water is the **solvent** and the numerous molecules and ions dissolved in it are the **solutes.** The solutes include protons (H^+), ions like sodium (Na^+), potassium (K^+), calcium (Ca^{2+}), organic molecules such as sucrose ($C_{12}H_{22}O_{11}$), polar and nonpolar molecules, and a host of other substances, the chemical nature of which determines the ease or difficulty with which they move across membranes.

Diffusion

All molecules possess **kinetic energy** and move in a random fashion. In solutions, solutes become distributed uniformly as they diffuse and occupy all available space. **Diffusion** is the net movement of a substance from a region of its higher concentration to a region of its lower as a result of the random movement of its individual molecules; or, in other terms, *down a concentration gradient.* The greater (steeper) the concentration gradient, the faster the movement. If nothing intervenes, the movement will continue until the concentration gradient is eliminated, i.e. until the substance is uniformly distributed. Most movement of materials in cells is by diffusion although it is neither the most efficient means nor can it be used for long distance moves.

Osmosis

Osmosis is a special kind of diffusion that pertains specifically to water: the movement of water across a selectively permeable membrane that permits the passage of water but inhibits the movement of the solute. The water moves down a concentration gradient from the region of its higher concentration of free water molecules (less solutes) to the region of its lower concentration of free water molecules (more solutes), or from high pressure to low pressure.

In comparing the relationship of the cell contents to those of the surroundings, three terms are used: 1.) **isotonic:** The two solutions have the same concentration of solutes, hence the same amount of water moves into the cell as moves out; 2.) **hypotonic:** The water outside the cell has *less* solute (hypo = less), and therefore *more* free water with the result that water moves into the cell at a greater rate than it moves out; 3.) **hypertonic:** The water outside the cell has *more* solute (hyper = more), and therefore *less* free water with the result that water moves out of the cell at a greater rate than it moves in.

In osmosis, water moves from a hypotonic solution to a hypertonic through a selectively permeable membrane. Water will diffuse across a selectively permeable membrane until the concentrations are the same on both sides (i.e. isotonic). If pressure is applied to the hypertonic side (the side into which the water is moving), it is possible to stop the inward flow of water. The amount of pressure needed to do so is called the **osmotic pressure** of the solution and is determined by the concentration of total solutes in the solution. Osmosis doesn't depend on the *kinds* of molecules or ions in solution, only on the *amount* of solutes.

Osmosis is vitally important for plants because it enables the plant to assimilate nutrients from the soil; the soil water is hypotonic to the root cells. Osmosis also makes the cells **turgid** (swollen) and gives rigidity to the plant. Water in the cell (mostly in the central vacuole) exerts a **turgor pressure** against the cell wall, which, in turn, exerts inwardly a mechanical **wall pressure** against the protoplast. The two equal and opposing pressures give strength to the cell and columns of water-filled cells keep the plant erect. Forget to water a house plant and the cells lose water, the turgor and wall pressures lessen, the cells become **flaccid** (limp), and the whole plant **wilts.** Internally, as water leaves the cells the cytoplasm shrinks away from the wall and collapses into an interior clump; the cell is **plasmolyzed,** and the process is **plasmolysis** (an example of osmosis in action). The cells are not dead, but they stop active metabolizing. The wilted celery stalk retains more of its rigidity than the lettuce leaves because it has reinforcing strings of collenchyma cells among its thin-walled parenchyma. Wash off the salt solution and immerse the salad greens in pure water and if the membranes were not broken, osmosis will rehydrate the cells to turgidity.

The Structure of Membranes

The **plasma membrane** on the outer surface of the protoplast regulates what enters and leaves the cell. Other membranes within the cell compartmentalize the protoplast, separating the interior into units of differing chemical composition, each with their own **differentially (selectively) permeable membranes**—membranes, like the plasma membrane, that selectively permit the passage of some materials while inhibiting the passage of others. (A **freely permeable membrane** that allowed everything to pass would be useless for the cell as would an **impermeable membrane** that allowed nothing to pass.)

The cell is thus a collection of "factories" that import, manufacture, and export metabolic substances and that are separated from one another by membranes.

The membranes are channels of communication within and between cells and carry the directions supplied by the genes in the nucleus. Signals transmitted via specific molecules that institute reactions in their receptors organize the individual cells into an integrated multicellular organism. All of the cellular membranes are essentially alike in structure, but differ in their constituent proteins. Consequently, they vary in the types of materials they recognize and in the reactions they initiate and conduct.

The currently acceptable version of membrane structure is the **fluid-mosaic model,** which visualizes the membrane as a double layer of *fluid* lipids in which proteins float laterally, forming a changing *mosaic* pattern (refer to Figure 9-1).

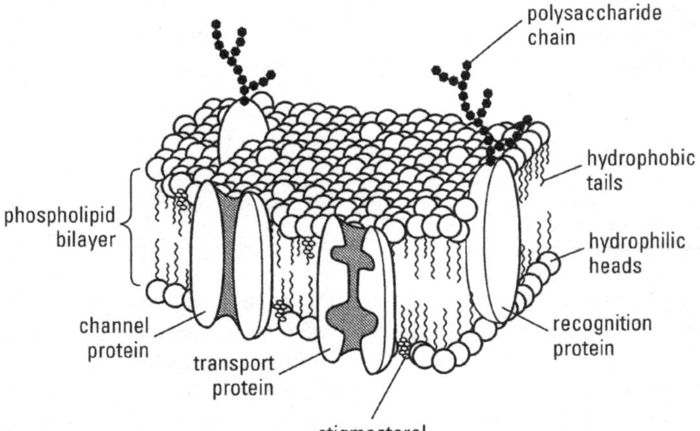

Figure 9-1

The **lipid bilayer** is formed primarily of **phospholipids,** which, in a watery solution, orient with their **hydrophilic** (water-loving) **heads** toward the outside and their **hydrophobic** (water-hating) **tails** to the inside. Other lipids, the **sterols,** are present in small numbers. (In animals, the representative sterol is cholesterol; in plants, **stigmasterol.**)

Embedded in this bilayer are proteins that function in three cellular activities: transport, reception, and communication. These proteins are structured with two different configurations: Some extend through the lipid layer, with their hydrophobic portions amongst the lipid tails and their hydrophilic portions protruding on either side of the lipid layer. These are the **transmembrane proteins.** The second group, the **peripheral proteins,** are mostly **glycoproteins** (proteins with short carbohydrate side chains) attached to the projecting ends of the transmembrane proteins. Some of the proteins are not attached and appear to float; others, the **integral proteins,** are firmly embedded and don't move.

Transportation of Materials Across Membranes

Concentration gradients by which diffusion and osmosis operate are only partially effective when all needs of the plant are considered. Some essential materials, for example, are present in small amounts in the soil but used in greater concentrations by the plant. How do the plants retain such materials as they accumulate in cells against the concentration gradient? Or, how is it possible for plants to get rid of unneeded or toxic substances that diffuse inward? Three means of transport take place: passive, active, and vesicle.

Passive transport

Simple diffusion. Some materials, like water, simply diffuse across the cell's membrane as a function of the concentration gradient across the membrane, i.e. they move from a region of their higher concentration to a region of their lower concentration through the phospholipid layer. Materials such as dissolved gases—oxygen and carbon dioxide, for example—also diffuse across, as do lipid-soluble substances. Sufficiently small and nonpolar molecules (like O_2) or small and unchaged polar (like H_2O) move readily if a concentration gradient exists. The rate of movement depends upon the steepness of the concentration gradient. The energy for simple diffusion is the kinetic energy inherent in all substances.

Facilitated diffusion. Charged molecules and polar molecules pass through membranes using protein channels. Like simple diffusion, these channels depend upon concentration gradients between the two sides of the membrane, but unlike simple diffusion, use **carrier proteins** and **channel proteins** to assist in the movement of materials. Carrier proteins chemically bind with the moving material, passing it from one binding site to another, literally carrying it through the lipids of the membrane. The channel proteins, on the other hand, open a water-filled passage/channel through which ions move, bouncing from one temporary binding site to another down the channel. The channels are gated—they open and close. Some of the channels appear to facilitate the movement of water exclusively and use channel proteins called **aquaporins**.

Active transport

Moving against a concentration gradient or an electrochemical gradient requires the input of additional energy and so is called the **active transport** of materials. This is the only kind of transport able to move molecules against a concentration gradient. Like facilitated transport, it depends upon protein transporters in the membrane; in this process these proteins are composed of the membrane-bound enzyme, H^+-ATPase. Using the typical proton pump reaction, the enzyme catalyzes the hydrolysis of ATP. In primary active transport, the energy released by the hydrolysis is used to modify the membrane protein itself, which then transports the molecule through the protein. In secondary active transport, the energy is used to pump a large quantity of H^+ across the membrane; as the protons flow back they carry with them other substances. Sucrose is moved in plants similarly by a proton-sucrose cotransport system.

Vesicle transport

Some materials that move among cells can't be transported in any of the previous ways, but must be transported by enclosing and then moving them in **vesicles** (small balloon-like structures). Large proteins and polysaccharides are commonly moved this way; the materials for cell walls, for example, are carried to the construction site in vesicles pinched from the endoplasmic reticulum as are the enzymes used to digest insects captured by carnivorous plants. The process in general is called **exocytosis** because it conducts materials away (exo-) from the cell (cyto-).

Some materials enclosed within vesicles move into the cell in a process of **endocytosis** (endo meaning within). Once thought to occur only in animals or animal-like protists, it now is understood to be a process of plants as well where one application carries stray pieces of ready-made plasma membrane back for recycling.

Reception and Communication Functions of Membranes

Some of the proteins embedded in the phospholipid bilayer are receptors that receive a signal molecule. This molecule contains a message that is transmitted to another molecule, usually with instructions to perform some action. Some of the signal molecules pass through the membrane in ways described above, but many rely on membrane receptors to conduct the instructions to secondary messengers within the cell, a process of **signal transduction** (changing the molecular form of the signal). The secondary messengers then pass the transduced signal to the activator cells, which carry out the function.

The calcium ion (Ca^{2+}) is the principal second messenger in plant cells and is stored in the vacuole. It moves out through special calcium channels in the tonoplast when a signal is received on the membrane and in the cytosol binds to calmodulin. (Some plants store Ca^{2+} in the lumen of the endoplasmic reticulum.) The Ca^{2+}-calmodulin then activates enzymes in the membranes.

Auxins (the plant growth substances similar to the hormones of animals) have several receptors on the membranes, each of which activates a different set of reactions. Receptors receive signals from other organisms also. These signals trigger interactions between plants as hosts or as prey to fungi, bacteria, protists, or to other species of plants. A group of receptors named **lectins** are scattered throughout the membrane system. They are proteins and glycoproteins that recognize, among other things, the specific bacteria that form root nodules with legumes in the nitrogen-fixing symbiosis. Other lectins are involved in recognition of the pollen that falls on the stigmas of flowers. Pollen tubes are initiated only if pollen and stigma are of compatible species.

CHAPTER 10
RESPIRATION

Respiration Releases Energy for Plant Metabolism

Respiration is the process through which energy stored in organic molecules is released to do metabolic work. A stepwise process conducted in all living cells, it is controlled by enzymes, and releases carbon dioxide and water.

$$\underset{\text{glucose}}{C_6H_{12}O_6} + \underset{\text{oxygen}}{6O_2} \longrightarrow \underset{\text{carbon dioxide}}{6CO_2} + \underset{\text{water}}{6H_2O} + \text{energy}$$

Breathing, the inspiration and expiration of air by animals, is not the same as respiration. Both animals and plants respire, but plants neither breathe nor have specialized respiratory systems as do animals. In plants, gases diffuse passively into the plant (through the stomata or directly into the epidermal cells) where they come into contact with the moist cellular membranes and then move in water along diffusion gradients between and within cells. No special carriers (such as the hemoglobin of human blood) or organs (such as lungs or gills) aid in the diffusion.

Glucose is the originating molecule for respiration; other reserve foods either follow different utilization pathways or, in the case of complex carbohydrates, are broken down to glucose before undergoing respiratory oxidation.

Respiration can be divided into the following stages (see Figure 10-1):

- **Glycolysis** is the breakdown of a 6-carbon glucose molecule into two molecules of 3-carbon pyruvate; it takes place in the cytoplasm of *all* living cells.

- If oxygen is present (**aerobic respiration**), pyruvate is used in the following reactions that take place in the mitochondria:
  - The **Krebs cycle** (citric acid cycle)occurs in the matrix.
  - **Electron transport chain** and **oxidative phosphorylation** occur deep in the cristae.
- If oxygen is not present (**anaerobic respiration**), pyruvate is used in **fermentation**.
  - **Lactate formation** occurs in in animal, bacteria, fungi, and protist cells.
  - **Alcohol fermentation** occurs in in yeast and plant cells.

The thermodynamic efficiency—the percentage of the potential energy of the glucose molecule that is recovered to do work in the cells—varies between 22–38 percent in aerobic respiration and is considerably less in anaerobic. (Gasoline engines average less than 25 percent efficiency.) The lost energy is released as heat, some of which is used by plants in interesting ways.

The first steps of energy release (glycolysis) in all organisms follow the same general pattern. This pattern presumably originated early on Earth with the single-celled prokaryotes before molecular oxygen was plentiful in the atmosphere and before the advent of cellular organelles. Only after photosynthesis altered the gaseous content of the atmosphere could the second chain of respiratory reactions develop in which oxygen is used as an electron acceptor. Some of the small, obligate anaerobes today still respire exclusively with the glycolytic pathway, but most larger organisms resort to glycolysis only for short periods when oxygen is temporarily absent (plant roots in flooded soils, for example) or oxygen can't get to cells fast enough (such as when human muscles are worked hard during exercise).

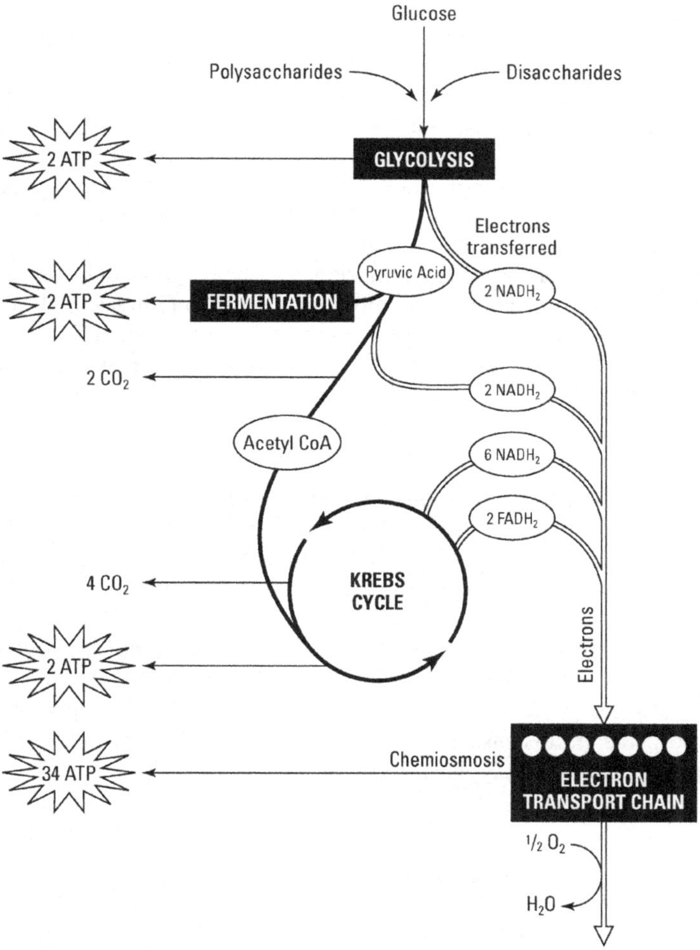
Glucose
Polysaccharides
Disaccharides
2 ATP
GLYCOLYSIS
Electrons transferred
Pyruvic Acid
2 NADH2
2 ATP
FERMENTATION
2 CO2
2 NADH2
Acetyl CoA
6 NADH2
2 FADH2
4 CO2
KREBS CYCLE
Electrons
2 ATP
34 ATP
Chemiosmosis
ELECTRON TRANSPORT CHAIN
1/2 O2
H2O

Figure 10-1

Glycolysis

Simply put, glycolysis (glyco = sugar; lysis = splitting) splits a 6-carbon sugar, glucose, into two molecules of 3-carbon pyruvate in a series of steps, each catalyzed by a particular enzyme. The energy of the sugar molecule is released in stages and transferred to adenosine triphosphate (ATP) and the reduced coenzyme nicotinamide adenine dinucleotide (NADH), each of which are then available for other energy-requiring metabolic processes. (Note: Pyruvic acid formed in the reactions dissociates to pyruvate plus H^+; the acid and the pyruvate exist together in equilibrium, and so the names "pyruvate" and "pyruvic acid" are used interchangeably by textbook authors.)

First steps (preparatory, energy-using stages)
The energy needed to start the reaction comes from an ATP molecule that is added, together with a phosphate group, to one of the sugar carbons, thereby energizing the sugar to glucose-6-phosphate.

The molecule is restructured, a second ATP enters, binding another phosphate group to a different carbon atom, and the 6-carbon energized sugar molecule splits into two 3-carbon molecules, each with a phosphate group consisting of dihydroxyacetone phosphate (DHAP) and glyceraldehyde 3-phosphate (PGAL).

Final steps (energy-releasing phase)
DHAP converts to PGAL. There are now *two* molecules of PGAL going forward. Two almost simultaneous reactions occur next: an oxidation and a phosphorylation. Energy is harvested in the oxidation of PGAL (H atoms and their electrons are removed), and the coenzyme NAD^+ is reduced (H atoms and electrons added) to form NADH, a high-energy compound.

The phosphorylation, using some of the oxidation energy, attaches an inorganic phosphate group (P_i) to the carbon ring (the bond energy is similar to that of the high energy bonds of ATP). The substrate is now 1,3-bisphosphoglycerate, with two phosphate groups attached.

A phosphate group is released to phosphorylate a molecule of ADP, *forming a molecule of ATP*. This kind of phosphorylation—formation of ATP by transferring a P_i from a metabolic intermediate compound—is called **substrate-level phosphorylation**. Since there are two PGAL molecules from each glucose molecule, two ATPs are formed at this stage. Two ATPs were needed to activate the pathway, so at this point the energy use and production are equal.

The substrate molecule is restructured: The remaining P_i group is moved to another binding site in the molecule, and a molecule of water is removed resulting in a high-energy, phosphorylated compound being formed (phosphoenolpyruvate).

A second substrate-level phosphorylation takes place; the P_i is moved to ADP, and the second molecule of ATP is formed.

The result of all of these reactions is: two molecules of 3-carbon pyruvate, four ATPs, and two NADHs for each molecule of glucose. The *net* gain of ATP, however, is only two molecules of ATP since two molecules were used to initiate the process. The two pyruvate molecules retain about 80 percent of the energy of the glucose molecule, and the two NADH molecules also are used in later energy-requiring metabolic reactions in the mitochondria. (Six molecules of ATP can be produced by re-oxidizing them back to NAD^+.)

Remember that the above reactions take place in the cytosol of cells and are catalyzed by enzymes specific for each reaction. Note that oxygen is not used in any of the above reactions. Glycolysis is an anaerobic process.

$$\text{glucose} + 2\text{NAD}^+ + 2\text{ADP} + 2\text{P}_i \longrightarrow$$
$$2\text{pyruvate} + 2\text{NADH} + 2\text{H}^+ + 2\text{ATP} + 2\text{H}_2\text{O}$$

Aerobic Respiration

The Krebs Cycle (Citric Acid Cycle)

If oxygen is present, the pyruvate produced in glycolysis is oxidized further in the Krebs cycle, a continuation of aerobic respiration. The cycle involves a series of changes in organic acids with 4-, 5-, and 6-carbon atoms (citric acid has 6 carbons). The cycle takes place in the mitochondria, separated from the reactions of glycolysis, which occur within the cytoplasm. The cycle starts (and ends) with acetyl coenzyme A (**acetyl CoA**) as the substrate. As the cycle turns, NADH, ATP, $FADH_2$ and CO_2 are released and the substrate is regenerated to be used again. Although carbon dioxide—an end product of respiration—is generated and the glucose completely oxidized during the cycling, this is not the final stage in aerobic respiration; most of the energy still remains in the high-energy electron carriers NADH and $FADH_2$.

Role of the mitochondria

A mitochondrion has two membranes, an outer and an inner, the latter deeply creased and folded thereby increasing its surface area on which the specialized reactions occur. The folds of the mitochondrion are called **cristae**. The inner cavity—the **matrix**—is filled with a watery solution containing all but one of the needed materials of the Krebs cycle. The electron carriers associated with the uptake of oxygen (used in the next step of respiration, see below) are embedded in the cristae.

Steps in the cycle

Preparation of pyruvate. The pyruvate produced in glycolysis is transported to the matrix of the mitochondria where it is converted to acetyl-CoA, a three-step conversion.

- The two 3-carbon pyruvate molecules are oxidized to two 2-carbon acetyl groups (CH_3CO).
- Two molecules of CO_2 are released.
- Two molecules of NADH are formed.

The acetyl groups are then attached to molecules of coenzyme A, forming acetyl-CoA, and the cycle begins.

Cyclical reactions. Acetyl-CoA, the substrate starting point, releases its 2-carbon portion to the 4-carbon organic acid, oxaloacetic acid, making the 6-carbon citric acid. The seven steps are repeated and the cycle is back to oxaloacetic acid—in the process having lost two carbon molecules to two molecules of CO_2 and releasing electrons to the high-energy compounds NADH, ATP, and FADH. Two turns of the cycle are necessary to release the six carbons of the original glucose molecule.

The energy input into the Krebs cycle comes from the oxidation (i.e. the removal of electrons) of NADH and $FADH_2$ and their regeneration back to NAD^+ and FAD. Without a constant supply of the latter compounds, the Krebs cycle stops.

$$\text{oxaloacetate} + 2CO_2 + \text{CoA} + \text{ATP} + 3\text{NADH} + 3H^+ + FADH_2 \longrightarrow \text{oxaloacetate} + \text{acetyl CoA} + 3H_2O + \text{ADP} + P_i + 3NAD^+ + \text{FAD}$$

The Electron Transport Chain and Phosphorylation

After the Krebs cycle is completed, oxygen enters the respiration pathway as the electron acceptor at the end of the electron transport chain.

The oxidation takes place in a series of steps, like the electron chain of photosynthesis, but with different transport molecules. Many of the latter are **cytochromes** (proteins with an iron-containing porphyrin ring attached) where the electron exchanges take place on the iron atoms. Others are iron-sulfur proteins with iron again the exchange site. Three complexes of carriers are embedded together with proteins in the inner mitochondrial membrane where they assist in the **chemiosmotic** production of ATP (see below). The most abundant electron carrier, **coenzyme Q (CoQ)**, carries electrons and hydrogen atoms between the others.

The transport chain often is likened to a series of magnets, each stronger than the last, which pull electrons from one weaker carrier and release it to the next stronger one. The last acceptor in the line is oxygen, an atom of which accepts two energy-depleted electrons and two hydrogen ions (protons) and forms a molecule of water.

Energy from the transport chain establishes a proton gradient across the inner membrane of the mitochondrion and supplies the energy for the embedded protein complexes—which also are proton pumps and sites of the chemiosmotic process. As electrons are pulled from NADH and $FADH_2$, protons (H^+) also are released, and the protein complexes pump them into the intermembrane space. Since the membrane is impermeable to protons, they accumulate there, and thus both a H^+ gradient and an electrochemical gradient are established between the inner membrane space and the matrix. Embedded in the membrane, however, are complexes of the enzyme **ATP synthase** with inner channels through which protons can pass. As the protons move down the gradient, their energy binds a phosphate group to ADP, an oxidative phosphorylation, making ATP.

The importance of the Krebs cycle and oxidative phosphorylation is evident when the net yield of ATP molecules produced from each molecule of glucose is calculated. Each turn of the Krebs cycle produces one ATP, three molecules of NADH, and one of $FADH_2$. (Remember that it takes *two* turns of the cycle to release the six carbons of glucose as CO_2 so this number is doubled for the final count.) The retrieval of energy from the oxidative phosphorylations and the chemiosmotic pumping are an impressive 34 ATPs (four from the two NADH molecules produced in glycolysis and added to the transport and phosphorylation chain; six from the NADH molecule produced in the conversion of pyruvate to acetyl CoA; and 18 from the six molecules of NADH, four from the two FADH molecules, and two directly produced in two turns of the Krebs cycle.) The net yield from glycolysis is only two ATP molecules.

The number of enzymes and the precise mechanisms of the respiratory pathways may seem to be an unnecessarily complex way for cells to obtain energy for metabolic work. But, if electrons were added directly to oxygen, the reaction probably would produce enough heat to damage the cells and result in too small an amount of captured energy to be a significant source for future energy needs.

Anaerobic Respiration: Fermentation

A supply and demand problem arises among cells when glycolysis produces more NADH than can be utilized or when NAD^+ supplies are diminished or oxygen is unavailable. NADH production in glycolysis is a way to dispose of electrons and hydrogen; the NADH needs the electron transport chain with its terminal oxygen acceptor and NAD^+ is needed to complete the conversion of PGAL to pyruvate. If the pathway is disturbed, organisms remedy the problem generally in one of two ways.

Lactate fermentation

$$\text{glucose} + 2\text{ADP} + 2\text{P}_i \longrightarrow 2\text{lactate} + 2\text{ATP} + 2\text{H}_2\text{O}$$

Animals, protists, and many bacteria and fungi make lactate and release two molecules of ATP, enough to regenerate some NAD^+ and keep glycolysis running (but utilizing only a small portion of the energy of the glucose). Yogurt and cheese makers employ bacteria that respire this way and harvest the tasty byproducts of the reactions.

Alcohol fermentation

$$\text{glucose} + 2\text{ADP} + 2\text{P}_i \longrightarrow 2\text{ethanol} + 2\text{CO}_2 + 2\text{ATP} + 2\text{H}_2\text{O}$$

Most plant cells and yeasts (fungi) breakdown pyruvate to acetaldehyde, releasing CO_2. The acetaldehyde is then reduced by NADH to ethanol (ethyl alcohol). The CO_2 makes bread rise, and ethanol is used by brewers and distillers to make alcoholic beverages of all kinds.

Thermodynamically, this is a poor use of glucose. Over 90 percent of the energy of glucose remains in the two alcohol molecules; fermentation has removed only about 7 percent. The ATP captures about one quarter of that, with the rest released as heat.

CHAPTER 11
PHOTOSYNTHESIS

The Most Important Process in the World

Photosynthesis is the process by which organisms convert light energy into chemical energy. The most common and critical type of photosynthesis takes place in chlorophyll-containing **plants, algae,** and **cyanobacteria.** These organisms capture radiant energy of the sun and, by utilizing carbon dioxide and water, convert it to chemical energy stored in molecules of carbohydrates. Oxygen and water are released as by-products. The generalized equation is:

$$6CO_2 + H_2O = C_6H_{12}O_6 + 6O_2 + 6H_2O$$

carbon dioxide + water = glucose + oxygen + water

The other photosynthesizers are **green** and **purple bacteria,** which convert light energy to chemical energy, but use other raw materials and pigments in an oxygen-free (**anaerobic**) environment to make carbohydrates. Different by-products result.

A generalized equation for *all* kinds of photosynthesis therefore is:

$$nCO_2 + 2nH_2X = nCH_2O + 2nX + nH_2O$$

carbon dioxide + hydrogen = carbohydrate + byproduct + water

Photosynthesis is the single most important process on earth, for without it neither plants nor animals (including humans) could survive. The energy stored in organic molecules by photosynthesizers is the fuel of life for most living things and the oxygen released during photosynthesis makes cellular respiration—and therefore life—possible on earth.

Overview of Eukaryote Photosynthesis

Photosynthesis in **plants** and **algae** takes place in **chloroplasts** and entails **two steps:**

1. **Energy transferring (energy-transduction) reactions** (commonly called the light-dependent or light reactions)
2. **Carbon fixation reactions** (sometimes inappropriately called the dark reactions)

Step one: Energy transfer

The energy transferring reactions are photochemical processes that take place in two physically separate but chemically linked **photosystems: Photosystem I (PsI)** and **Photosystem II (PsII).** Photosystems are pigment molecules that capture energy from the sun and are arranged in the thylakoid membranes of the chloroplasts. The chlorophyll and other pigments of both photosystems absorb light energy, most of which is stored temporarily in energy-rich chemical bonds of **ATP** (adenosine triphosphate) and the electron carrier **NADPH** (reduced nicotinamide adenine dinucleotide phosphate). ATP and NADPH supply the energy for the resultant carbon fixation reactions of step two. Oxygen (O_2) is a by-product of water molecules splitting in the initial energy exchanges of step one. The three products of the energy transfer phase are ATP, NADPH, and O_2.

Step two: Carbon fixation

The carbon fixation reactions of the second step of photosynthesis are biochemical and use the energy of ATP and the reducing power of NADPH to repackage the energy in a form that can be transported and stored, as the carbohydrates sugar and starch. Carbon fixation reactions do not require light; if cellular energy is available, the reactions occur.

Plants have developed three different pathways for photosynthetic carbon fixation, one basic procedure and two modifications of it.

- **C3 Pathway** (also called the **Calvin cycle** after its 1961 Nobel Prize-winning discoverer). This method is used by most common temperate zone species.
- **C4** or **Hatch-Slack Pathway.** An additional step is added to the Calvin cycle, making it more efficient for plants structurally modified to do so. Many common grasses and tropical plants use this pathway; it is a necessary adaptation in areas of high light intensity, high temperatures or semi-aridity.
- **CAM (crassulacean acid metabolism) Pathway.** Another Calvin cycle modification is made by succulents and other plants growing in areas of high temperatures, high light, and low moisture (deserts especially). In this modification, carbon fixation takes place at night in a pathway similar to C_4 photosynthesis and, in addition, during the day carbon is fixed in the same cells using the C_3 pathway. This pathway is named for the family of plants, Crassulaceae, in which it was first discovered.

Products

The final products of carbon fixation are a disaccharide sugar, **sucrose,** and a polysaccharide, **starch**. The sucrose is formed from two monosaccharides (6-carbon or hexose sugars), **glucose** and **fructose,** joined together by an extra oxygen atom. Stored energy is transported from cell to cell in plants by the water-soluble sucrose. (In vertebrates, glucose is the transported sugar.)

Starch molecules are strings of glucose molecules too large to move through membranes, and, therefore, useful for storing energy. As energy is needed, the starch is converted to sucrose and transported. Plants build and fuel their bodies from these carbohydrates.

Two intermediate carbohydrates (manufactured before sucrose or starch) are the first detectable products in the C_3 and C_4 Pathways. In the C_3 Pathway the product is **PGA (3-phosphoglycerate)** (3 carbons), and in C_4 photosynthesis the first detectable product is **oxaloacetate** (4-carbons).

Details of Photosynthesis in Plants

Photosynthesis is a complex of interactions taking place at special times and sites and with special materials, but relying upon many standard metabolic procedures used elsewhere in plants and other organisms. All of the reactions are catalyzed (promoted) by specific enzymes.

The description of photosynthesis here refers to structures and actions in plants, but the fundamentals apply as well to algae and to some aspects of procaryote (cyanobacterial) photosynthesis. Most of the reactions occur simultaneously in nanoseconds (10^{9}) or less in various parts of the chloroplasts, but understanding the process is easier if it's separated into sequential steps. The overall process is shown in Figure 11-1.

Energy transferring reactions

The pigment molecules that capture energy from the sun are arranged in the thylakoid membranes of the chloroplasts in structurally separate units called **photosystems.** Hundreds of systems are present in each thylakoid. Each photosystem has a light collecting array of 200–300 molecules, the **antenna complex,** not unlike a satellite dish, which collects and focuses the photons into the **reaction center** where energy processing begins. The photons move from antenna molecule to antenna molecule without an energy loss, a resonance energy transfer.

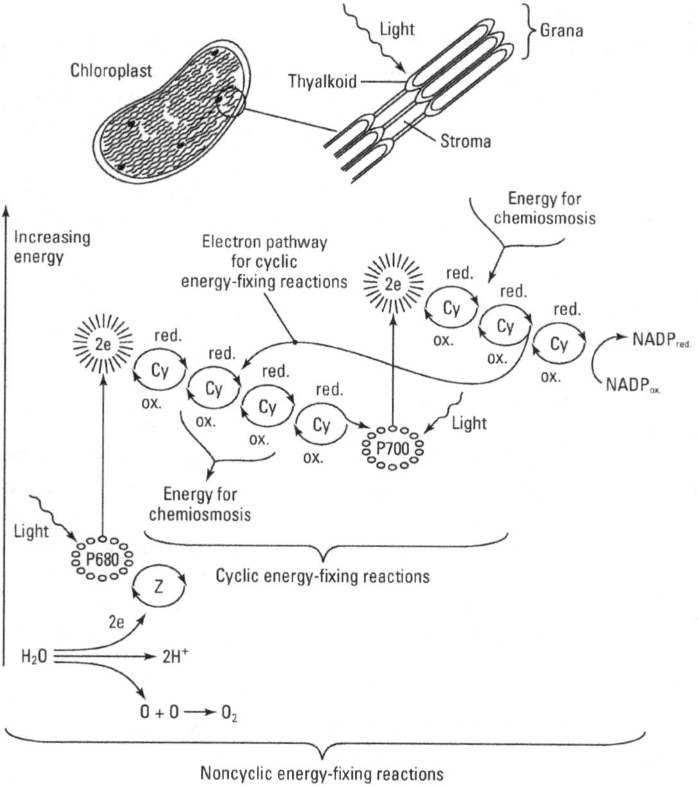

Figure 11-1

There are two kinds of photosystems, **Photosystem I (PsI)** and **Photosystem II (PsII).** In the reaction center of PsI the light-absorbing pigment is a specialized chlorophyll *a* molecule that absorbs red light of 700 nanometer wavelength most efficiently, hence the designation **P700.** PsII reaction center chlorophyll *a* molecules absorb maximally at 680 nanometers and therefore are **P680** molecules. Why *two* systems? Apparently because one alone can't capture enough energy to power the carbon-fixation reactions and to supply the rest of the energy requirements of plant metabolism. By utilizing chlorophylls and accessory pigments with different **absorption spectra** and providing different mechanisms to change the light energy to chemical energy, plants meet their energy needs.

PsI and PsII are chemically linked. PsI was discovered and named before PsII was identified. The process, however, starts with PsII, not PsI, and works as follows.

Photosystem II actions. The energy absorbed by a P680 molecule of chlorophyll in the reaction center of PsII converts the molecule to its excited state and raises the energy level of an electron, which moves to an outer orbit and is lost from the molecule. An acceptor molecule in the adjacent **electron transport system**—a chain of alternately oxidized and reduced compounds—accepts it, and immediately moves it down the transport chain, losing some energy in each transfer. The P680 molecule, having lost an electron, is unstable; it returns to its stable ground state by drawing an electron from a water molecule using an as yet not clearly understood mechanism of **photolysis.** The P680 is then ready to accept another photon, and the process is repeated over and over.

The energy lost in the electron chain transfers is used to produce **ATP** (adenosine triphosphate), the molecule universally used by organisms as a quick energy source. To do so, **proton pumps** in the thylakoid membrane are activated by the high-energy electrons. They pump protons (H^+) into the intermembrane space creating a (H^+) gradient across the thylakoid membrane. As protons diffuse back down the gradient into the stroma (the space in the chloroplast), they pass through pores lined with **ATPsynthase,** the enzyme that catalyzes the formation of ATP from ADP (adenosine diphosphate). In the passage, a phosphate group (P_i) is added to ADP forming a terminal high energy bond and ATP. The general term for this process is **photophosphorylation** (*photo* meaning light energy; *phosphorylation* indicating a phosphate group is added to an organic molecule).

The energy that reaches the end of the electron transport chain is transferred to the P700 chlorophyll in the reaction center of PsI where it replaces the electrons lost by P700 to the PsI electron transport chain.

Photosystem I actions. PsI is structurally similar to PsII and does part of the same things: It captures light energy (at a slightly different wavelength, 700 nm); transfers electrons down an electron chain; and captures the lost energy and uses it for short-term energy release and to power the carbon-fixing reactions in the next step of photosynthesis.

The enzymes used and products made differ in the two systems. In PsI the primary acceptor molecule of the electron chain is a special kind of chlorophyll *a* (A_0). The energy transfers are extremely rapid, on the order of picoseconds (10^{-12}) to femtoseconds (10^{-15}), and almost 100 percent of the energy is captured in the transfer from P700 to A_0. At least 3 more transfers occur in rapid succession before the vital transfer is made to the coenzyme, **$NADP^+$** (nicotinamide adenine dinucleotide phosphate). With this transfer, $NADP^+$ acquires a negative charge and attracts a proton (H^+) thus forming **NADPH** (reduced nicotinamide adenine dinucleotide phosphate). NADPH is important in cellular metabolism because it has great reducing ability: it readily releases its hydrogen atom, which drives various chemical reactions. At this point in photosynthesis, NADPH supplies the H^+ that eventually reduces CO_2 to a carbohydrate.

Role of water. The photo-oxidation (**photolysis**) of water is a key reaction and the primary source of the earth's oxygen. The mechanism is not completely understood, but this much is clear: Electrons lost from the reaction center of PsII are replaced by electrons removed from water in a stepwise process. The P680 molecule removes electrons one at a time from an oxygen-evolving complex. When four have been removed from two water molecules, oxygen (O_2) is released together with 4 protons (H^+) and 4 electrons (e^-). The protons are added to the reservoir in the space between the thylakoid membranes and function in the proton pumps of ATP synthesis. Some of the oxygen is used in plant cell respiration (see "Photorespiration" later in this chapter); the rest is released to the atmosphere.

Cyclic and noncyclic photophosphorylation. The passage of electrons from water (PsII PsI $NADP^+$) is a **noncyclic electron** flow—electrons that reach $NADP^+$ are not returned to water. This passage is a phosphorylation process because ATP is generated along the way by the addition of a phosphate group (P_i) to ADP. Since solar energy runs the process, this is also a **noncyclic photophosphorylation.**

Cyclic photophosphorylation also occurs in PsI in a process that does *not* make NADPH, split water, or produce oxygen. Its sole end product is ATP (making it a photophosphorylation) and the electrons cycle, without involving PsII or $NADP^+$. Thus, this is a **cyclic photophosphorylation** and also a **cyclic electron flow.** Since NADPH is not produced there is no reducing power for carbon fixation. Cyclic photophosphorylation seems designed for compilation of energy only. It apparently was the type of photosynthesis used by primitive organisms and is the method used today by some bacteria. Eukaryotes apparently retain the cyclic process as an evolutionary artifact.

Carbon fixation reactions

Calvin cycle. This is the basic cycle of photosynthesis, and it consists of three stages:

1. **Fixation:** CO_2 diffuses into the *stroma* of the *chloroplast* where the ATP generated earlier in Step One is used to chemically bind CO_2 to **ribulose 1,5-bisphosphate (RuBP)**, a 5-carbon sugar. The 6-carbon molecule produced is unstable and immediately separates into two, 3-carbon molecules of *3-phosphoglycerate (PGA)* with the assistance of the enzyme **ribulose 1,5-bisphosphate carboxylase/oxygenase (rubisco).**

2. **Reduction:** PGA, in a two-step process, is reduced (using the NADPH produced during PsI) to **3-phosphoglyceraldehyde (PGAL).**

3. **Regeneration:** the PGAL molecule is rearranged and re-creates RuBP, the starting substance to which CO_2 binds.

To produce glucose, the cycle has to make *three* turns, introducing a total of 3 atoms of carbon, one at each turn. To fix each requires 3 ATP energy molecules and 2 NADPH reducing power molecules for a final total of 9 ATP (3 molecules x 3 turns of the cycle) and 6 NADPH (2 molecules x 3 turns of the cycle). The material that leaves the Calvin cycle is not the 6-carbon glucose, but is a 3-carbon (triose) sugar, **PGAL (glyceraldehyde 3-phosphate)**, much of which moves from the chloroplast into the cytosol (the cytoplasm of the cell). Here the molecules in a series of reactions build the final product, sucrose. Not all of the PGAL migrates into the cytosol, however. Some remains in the chloroplast where it is converted into **starch** as an energy reserve of the cell.

Photorespiration

Photorespiration, a type of respiration, occurs in many plants in bright light. It differs from cellular respiration, which occurs in the mitochondria, because it does *not* release energy, although it does use O_2 and release CO_2. Photorespiration occurs at the same time that photosynthesis is operating and uses some of the newly made carbohydrate for energy, thereby reducing the yield of photosynthesis by as much as 50 percent in some plants and conditions. The enzyme **rubisco** is responsible for this seeming anomaly of efficiency.

Rubisco doesn't recognize CO_2 specifically enough. When O_2 is abundant in the chloroplast (as it is when photosynthesis is operating) rubisco accepts the O_2 rather than CO_2 and catalyzes a series of different reactions, resulting in carbon being released and energy being expended for no net energy gain. The structure of the leaves adds to the problem: The stomata, which regulate both CO_2 and O_2 gas exchanges, regularly close during periods of high light and temperature. While water is thereby conserved, gas exchange is impeded; entrance of CO_2 into the leaf is stopped as is the outward flow of O_2, slowing photosynthesis, but adding more O_2 in leaf tissues.

Photorespiration apparently is an evolutionary vestige left over from the time when the atmosphere contained little oxygen; molecular changes in the structure of rubisco have not occurred.

C_4 pathway

Some plants, notably grasses of tropical origin, have developed a means of subverting photorespiration by using a method called the **Hatch-Slack** or **C_4 Pathway,** named for its discoverers and its first product, **oxaloacetic acid,** a 4-carbon organic acid. A structural modification of the leaves accompanies the biochemical differences in this system: The leaf veins are surrounded by a cylinder of enlarged **bundle sheath cells** around which there usually is another layer of modified mesophyll cells. In cross-section view the two appear as rings of modified tissue, which gives the term **Kranz anatomy** (from the German word for wreath) to the leaf condition. The bundle sheath cells, in addition, contain functional chloroplasts unlike the sheath cells of C_3 plants. The first part of the C_4 pathway takes place in the mesophyll cells, the second part—which is an ordinary Calvin cycle—takes place in the bundle sheath cells.

The CO_2 diffusing into the leaf reacts in the cytoplasm of the mesophyll cells with **phosphoenolpyruvate (PEP)** and its enzyme **PEP carboxylase.** The **oxaloacetate** formed is then reduced by NADPH to **malate** some of which is shunted into the bundle sheath cells. (Malate also can be converted to *aspartate*, an amino acid.) Malate is decarboxylated in the bundle sheath cells to **pyruvate** and CO_2. This CO_2 now enters the Calvin cycle where, as usual, it becomes fixed first as PGAL and from there either to sucrose or to starch. The pyruvate goes back to the mesophyll cells, reacts with ATP, and synthesizes more PEP.

C_4 photosynthesis has both benefits and drawbacks.

- C_4 plants aren't adversely affected by the normal atmospheric (and leaf tissue) ratios of high O_2:low CO_2 concentrations that favor photorespiration because they *produce* CO_2. Rubisco doesn't switch to O_2, and there is no wasteful photorespiration in C_4 plants.

- C_4 plants are better adapted than C_3 plants to grow in hot, dry sites since they preserve cellular water loss with closed stomata in the heat of the day, but still have sufficient stores of CO_2 within the plant to conduct worthwhile photosynthesis.
- Photosynthesis in C_4 plants, however, requires two to three times more ATP energy to fix a molecule of CO_2 than it does in C_3 plants because of the extra malate step.

The presence of C_4 species in at least 19 different flowering plant families is evidence that the process is advantageous. The method apparently has arisen independently many times as a solution to a vexing environmental problem. Corn and sugar cane are commonly cultivated C_4 plants.

Crassulacean acid metabolism (CAM)

Another adaptation of plants for life in arid environments is the CAM photosynthetic pathway. CAM plants use *both* C_3 and C_4 pathways, but unlike the C_4 plants—which use two different types of cells—the CAM plants use only mesophyll cells and separate the time at which each pathway is run in the same cells.

CAM plants close their stomata during the day, which prevents water from leaving, but also prevents CO_2 from entering. They open the stomata at night for gas exchange when humidity is higher and temperatures lower. As CO_2 diffuses in, it combines with PEP in the cytoplasm of the mesophyll cells, forming oxaloacetic acid. The oxaloacetic acid is then reduced to malate, which accumulates in the cell vacuoles. This is the C_4 pathway, run in the dark of night with its product, malate, migrating only to the vacuole of the cell in whose cytoplasm it was made.

With the dawning day and rising temperatures, the stomata close once more, and photosynthesis commences. The malate is now transported back into the cytoplasm and is decarboxylated to **pyruvate** and CO_2. The CO_2 enters the chloroplast and is picked up in the Calvin cycle. PGAL and then sucrose or starch are produced.

The CAM plants are successful inhabitants of warm, arid sites and include species of 23 or more flowering plant families as well as a few ferns. Many are fleshy succulents like the stonecrops (*Crassulaceae)* for which the C_4 type was named, as well as many cacti, agaves, and lilies.

CHAPTER 12
MINERAL NUTRITION

The Essential Elements

Over 95 percent of the dry weight of a flowering plant is made up of three elements—carbon, hydrogen, and oxygen—taken from the air and water. The remaining 5 percent of the dry weight comes from chemicals absorbed from the soil. Roots absorb the chemicals present in their surroundings, but only 14 of the elements absorbed are necessary for plant growth. These 14 elements, along with carbon, hydrogen, and oxygen, are called the **17 essential inorganic nutrients,** or **elements.** Some of the essentials are needed in larger amounts than others and are called the **macronutrients;** those needed in lesser amounts are the **micronutrients.** All elements are needed in specific amounts. Note that there is a dispute among plant physiologists concerning the role of nickel in plant nutrition. Since many physiologists exclude it as essential, in some textbooks, lists like the following consist of only 16 essential inorganic nutrients. The 17 are:

- Macronutrients absorbed from the air: oxygen, carbon, and hydrogen.
- Macronutrients absorbed from the soil: nitrogen, potassium, magnesium, phosphorus, calcium, and sulfur.
- Micronutrients from the soil: iron, boron, chlorine, manganese, zinc, copper, molybdenum, and nickel.

An element is essential if it: 1.) is required for normal growth and reproduction; 2.) can not be replaced by another element; 3.) can be shown to be part of a molecule clearly essential to the plant structure or metabolism.

Plants use elements in differing amounts and forms, some as cations, others as anions. Almost all elements are used in a variety of ways, such as as catalysts for enzymatic reactions (either as part of the enzyme structure or as regulators or activators), as regulators of the movement of water in or out of the cell and maintenance of turgor pressure, as regulators of membrane permeability, as structural components of the cell or of electron receptors in the electron transport system, or as buffers (which maintain the pH within cells).

Two-thirds of all the naturally occurring chemical elements have been found in plants. Some odd kinds are known to be used metabolically by particular species, but others with no known function are accumulated apparently because they are present in the soil from which the plant is extracting water and ions. These non-useful chemicals are sequestered in cell vacuoles, as crystals, or as non-soluble compounds and remain in the plant throughout its life. Plants, therefore, can be useful in locating deposits of minerals, e.g. gold or uranium, and have been used by modern prospectors who collect the vegetation from a site and run spectroscopic analyses on the tissues. Some plants grow only in soils in which a particular element is present and are said to be **indicator plants** of that element.

Table 12-1 highlights the roles of the essential elements in plant nutrition.

The Role of Soils

Both the physical and chemical structure of soil influences the ways that plants obtain minerals from it. Soil from an ecological standpoint is that part of the Earth's surface in which plants grow, including—the thin layer on the surface of rocks penetrated by mosses, as well as the rich, black loam underlying Iowa cornfields.

Table 12-1: The Role of Inorganic Elements in Plant Nutrition and Their Deficiency Symptoms

Element	*Form in which Absorbed*	*Important Roles/Functions*	*Deficiency Symptoms*
Macronutrients			
carbon	CO_2	major component of organic compounds; presence defines "organic"	rarely limiting as a nutrient; no specific symptoms
hydrogen	H_2O	major component of organic compounds	rarely limiting as a nutrient; no specific symptom
oxygen	H_2O, O_2	major component of organic compounds	rarely limiting as a nutrient; no specific symptoms
nitrogen	NO_3^-, NH_4^+	part of amino acids, proteins, nucleotides, nucleic acids, chlorophylls, coenzymes	chlorosis; severe cases: turn yellow, die; some plants turn purple as anthocyanins accumulate in vacuoles; nutrient most likely to be deficient in soil
potassium	K^+	involved in osmosis, ionic balance, opening and closing of stomata; activator of enzymes; necessary for starch formation	weak, spindly stems and roots; older leaves especially affected—mottled with dead spots along margins and dead tips; roots more susceptible to disease
calcium	Ca^{2+}	component of middle lamella of cell walls; enzyme cofactor; involved in membrane permeability; component of calmodulin (regulator of membrane and enzyme activities)	root and shoot tips die; young leaves and shoots most affected, die back at tips and margins first
phosphorus	H_2PO_4 or HPO_4^{2-}	component of ATP and ADP (essential energy-carrying compounds), nucleic acids, several essential coenzymes, phospholipids of membranes	stunted growth of whole plant; dark green color; antho cyanins accumulate in vacuoles giving purple tinge to leaves; second most-likely nutrient to be deficient in soil

(continued)

Table 12-1: The Role of Inorganic Elements in Plant Nutrition and Their Deficiency Symptoms *(continued)*

Element	*Form in which Absorbed*	*Important Roles/Functions*	*Deficiency Symptoms*
magnesium	Mg^{2+}	center of chlorophyll molecule; activator of many enzymes	leaf tips and margins turn upward on mostly older leaves; chlorosis, mottling, some dead spots and reddish color of leaves
sulfur	SO_4^{2-}	component of some amino acids, proteins, and coenzyme A; can be absorbed through stomata as gaseous SO_2	young leaves with chlorosis between the veins: sulfur is rarely limiting
Micronutrients			
iron	Fe^{2+} or Fe^{3+}	required for chlorophyll synthesis; component of cytochromes and nitrogenase (important in respiration and photosynthesis)	short, slender roots; chlorosis between the veins in leaves
zinc	Zn^{2+}	activator or component of several enzymes; involved in auxin synthesis, maintenance of ribosome structure	leaf size and internodal length much reduced; leafe margins deformed; chlorosis between veins, especially in older leaves
molybdenum	MoO_4^{2+}	required for nitrogen fixation and nitrate reduction (nitrate reductase)	chlorosis starting in older leaves and progressing to younger; death of interveinal areas and then of whole leaf
boron	$B(OH)_3$ or $B(OH)_4$	influences Ca^{2+} utilization, formation of nucleic acids, maintenance of membranes; essential for growth of pollen tubes	young tissues most affected; apical meristems die; root tips swollen and discolored; young leaves yellow at base, twisted
copper	Cu^{2} or Cu^{2+}	activator of enzymes, present in some; involved in oxidation-reduction	wilting and twisting of dark green young leaves; often with numerous dead spots on blades; copper is rarely deficient

manganese	Mn^{2+}	activator of enzymes, required for O_2 release in photosynthesis, integrity of the chloroplast membrane; electron transfers	interveinal chlorosis and dead spots; thylakoid membranes disintegrate
chlorine	Cl^-	involved in water balance (osmosis), ionic balance; probably essential in photosynthetic O_2-releasing reactions	leaves wilt; turn reddish bronze in color; chlorosis, dead spots; stunted roots with abnormal thickening near tips
nickel	Ni	essential part of enzyme in nitrogen metabolism	leaf tips with dead spots

Soils are dynamic and ever-changing. They have a structure that both influences, and in turn is influenced by, the plants that they support. The weathering of rocks over time produces the soils from which plants extract water and all but one of the nutrients they need—carbon being the sole exception. Soils, in addition, provide anchorage for plant roots and thus a means of support for the aerial shoots. Air in the pore spaces of soils provides roots with oxygen for respiration and is a depository for the carbon dioxide released in the process. Though soils change constantly as materials are added and removed, in general they have five components: inorganic mineral particles, decaying organic matter, living organisms, air, and water. These five factors, mixed in almost infinite variety and proportions, give rise to different kinds of soils.

The rocks that produce soil contain mineral elements that, for the most part, are locked in a crystalline matrix and unavailable to plants until physical and chemical weathering loosen the chemical bonds. The nutrients must be released into the soil water before plant roots can absorb them. Those nutrients that become soluble often are leached out of the root zone before roots or soil organisms can pick them up. The positively charged ions (**cations**) such as Na^+, Ca^{2+}, Mg^{2+}, and K^+, on the other hand, are readily adsorbed on the negatively charged surfaces of the soil colloids–and as easily removed. The CO_2 released in respiration speeds cation exchange because it forms carbonic acid (H_2CO_3) in the soil water. The acid, in turn, splits into carbonate (HCO^{3-}) and hydrogen ions (H^+). It is the hydrogen ions that dislodge the cations from the colloids and thus make them available for use by plants and microorganisms. The **cation exchange capacity** of a soil is an important determinant in how well (or poorly) plants grow.

Most of the negatively charged ions (**anions**), e.g. nitrate (NO^{3-}) and sulfate (SO_4^{2-}), leach out because they are not attached to the colloids. Some anions, e.g. phosphate ($H_2PO_4^-$), adsorb to iron or aluminum compounds in the soil and ultimately become available to plants.

CHAPTER 13
GROWTH OF PLANTS

Regulating Growth: Plant Hormones

Plant cells are in constant chemical communication with one another and with their environment. They recognize and respond to stimuli of many kinds, using a system of chemical messengers that receive and transmit the stimuli via ordinary body cells (unlike the highly specialized cells of animal nervous systems). Control of the plant system apparently resides in the genes of each cell, which are turned on and off by the chemical messages they receive. The response may be **stimulatory** (initiating cellular division and enlargement, for example) or **inhibitory** (such as stopping a metabolic process).

The chemical messengers are **hormones**, organic substances manufactured in small amounts in one tissue and usually transported to another where they initiate a response. (A few act in the tissues where they are produced.) The hormone molecule itself carries little information and produces a reaction only when it binds to appropriate receptor molecules at the response site.

Plants, in comparison to animals, have both fewer hormones and fewer kinds of responses. Plant hormones, however, usually act in combination, thus producing more varied responses than if acting individually. The same hormone also can produce different effects when acting in different tissues or in different concentrations in the same tissue. The developmental stage of the plant additionally determines what effects the hormone activates. Growth and development depend upon a successful coordination of the activities of hormones, not just the presence or absence of individual ones.

Types of Plant Hormones

There are five general classes of hormones: auxins, cytokinins, gibberellins, ethylene, and abscisic acid.

Auxins

An auxin, **indole-3-acetic acid (IAA),** was the first plant hormone identified. It is manufactured primarily in the shoot tips (in leaf primordia and young leaves), in embryos, and in parts of developing flowers and seeds. Its transport from cell to cell through the parenchyma surrounding the vascular tissues requires the expenditure of ATP energy. IAA moves in one direction only—that is, the movement is polar and, in this case, downward. Such downward movement in *shoots* is said to be **basipetal** movement, and in *roots* it is **acropetal.**

Auxins alone or in combination with other hormones are responsible for many aspects of plant growth. IAA in particular:

- Activates the differentiation of vascular tissue in the shoot apex and in calluses; initiates division of the vascular cambium in the spring; promotes growth of vascular tissue in healing of wounds.
- Activates cellular elongation by increasing the plasticity of the cell wall.
- Maintains apical dominance indirectly by stimulating the production of ethylene, which directly inhibits lateral bud growth.
- Activates a gene required for making a protein necessary for growth and other genes for the synthesis of wall materials made and secreted by dictyosomes.
- Promotes initiation and growth of adventitious roots in cuttings.

- Promotes the growth of many fruits (from auxin produced by the developing seeds).
- Suppresses the abscission (separation from the plant) of fruits and leaves (lowered production of auxin in the leaf is correlated with formation of the abscission layer).
- Inhibits most flowering (but promotes flowering of pineapples).
- Activates tropic responses.
- Controls aging and senescence, dormancy of seeds.

Synthetic auxins are extensively used as herbicides, the most widely known being **2,4-D** and the notorious **2,4,5-T,** which were used in a 1:1 combination as Agent Orange during the Vietnam War and sprayed over the Vietnam forests as a defoliant.

Cytokinins

Named because of their discovered role in cell division (cytokinesis), the cytokinins have a molecular structure similar to adenine. Naturally occurring **zeatin**, isolated first from corn (*Zea mays*), is the most active of the cytokinins. Cytokinins are found in sites of active cell division in plants—for example, in root tips, seeds, fruits, and leaves. They are transported in the xylem and work in the presence of auxin to promote cell division. Differing cytokinin:auxin ratios change the nature of organogenesis. If kinetin is high and auxin low, shoots are formed; if kinetin is low and auxin high, roots are formed. Lateral bud development, which is retarded by auxin, is promoted by cytokinins. Cytokinins also delay the senescence of leaves and promote the expansion of cotyledons.

Gibberellins

The gibberellins are widespread throughout the plant kingdom, and more than 75 have been isolated to date. Rather than giving each a specific name, the compounds are numbered—for example, **GA1,**

GA2, and so on. **Gibberellic acid three (GA3)** is the most widespread and most thoroughly studied. The gibberellins are especially abundant in seeds and young shoots where they control stem elongation by stimulating both cell division *and* elongation (auxin stimulates only cell elongation). The gibberellins are carried by the xylem and phloem. Numerous effects have been cataloged that involve about 15 or fewer of the gibberellic acids. The greater number with no known effects apparently are precursors to the active ones.

Experimentation with GA3 sprayed on genetically dwarf plants stimulates elongation of the dwarf plants to normal heights. Normal-height plants sprayed with GA3 become giants.

Ethylene

Ethylene is a simple gaseous hydrocarbon produced from an amino acid and appears in most plant tissues in large amounts when they are stressed. It diffuses from its site of origin into the air and affects surrounding plants as well. Large amounts ordinarily are produced by roots, senescing flowers, ripening fruits, and the apical meristem of shoots. Auxin increases ethylene production, as does ethylene itself—small amounts of ethylene initiate copious production of still more. Ethylene stimulates the ripening of fruit and initiates abscission of fruits and leaves. In monoecious plants (those with separate male and female flowers borne on the same plant), gibberellins and ethylene concentrations determine the sex of the flowers: Flower buds exposed to high concentrations of ethylene produce carpellate flowers, while gibberellins induce staminate ones.

Abscisic acid

Abscisic acid (ABA), despite its name, does not initiate abscission, although in the 1960s when it was named botanists thought that it did. It is synthesized in plastids from carotenoids and diffuses in all directions through vascular tissues and parenchyma. Its principal effect is inhibition of cell growth. ABA increases in developing seeds and promotes dormancy. If leaves experience water stress, ABA amounts increase immediately, causing the stomata to close.

Tropisms

Responsive growth movements toward or away from an external stimulus are called **tropisms.** If the plant movement is *toward* the stimulus, it is a **positive tropism;** *away from* the stimulus, a **negative tropism.**

Phototropism

The tropic response to unidirectional light is called **phototropism.** In general, shoots grow toward light and hence are positively phototropic; roots grow away from light and are negatively phototropic. Well-known and often-repeated experiments with oat seedlings have shown that the auxin IAA, which causes elongation of cells, migrates to the shaded side of oat coleoptiles. The subsequent differential growth on the two sides causes the coleoptiles to bend toward the light. Although green stems also bend and grow toward the light, in this case an IAA inhibitor prevents cells from elongating on the lighted side, while those on the shaded side continue to elongate; the stem bends toward the light as a consequence of the differential growth. Different wavelengths of light cause differing growth responses. The blue end of the spectrum—wavelengths less than 500 μm—is most effective in producing a growth response.

Gravitropism

Gravitropism is the plant response to gravity. The mechanism of how gravity is sensed by plants is as yet unexplained. None of the numerous hypotheses is fully adequate. Over the eons, plants probably developed several methods to cope with this environmental factor.

Shoots are negatively gravitropic, because they grow upward; roots are positively gravitropic—they grow downwards. IAA, calcium ions (Ca^{2+}), and possibly ABA are involved in instigating growth and curvature in many plants. Still to be proven is the long-held belief that starch grains migrating from upper to lower sides in the root cap of a horizontally held root initiate the growth response.

Thigmotropism

The growth response of a plant or a plant part to the touch of a solid object is called thigmotropism. Tendrils of climbing plants wrapping around a support is a common thigmotropic response, accomplished by cells on the side touching the support shortening and those on the opposite side elongating. IAA and ethylene are two hormones probably involved in the response.

Other Plant Movements

Nastic movements

Although anchored in place by root systems, plants move their organs in response to many kinds of external stimuli. These movements are called nastic movements and differ from tropic movements in that they are not directed toward or away from the stimulus. Movements triggered by touch, such as closing the traps of insectivorous plants, are called **thigmonastic** or **seismonastic** movements. The changing daily cycles of light and darkness produce "sleep" (**nyctinastic**) movements in leaves of many species. Most of the actual nastic movements can be explained by changes in the turgor pressure of specially located parenchyma cells after a stimulus has been received.

Thigmomorphogenesis

The growth response to generalized mechanical disturbances is called **thigmomorphogenesis.** Plants in their natural environment are subject to all manner of jarring, touching, and shaking by the wind, passing animals, rain, and the like. The general response of most plants to such disturbances results in decreased height, increased diameter, and more supportive tissues in the shoots. The change in the general form of the plant apparently results from activation of genes, one of which carries the code for **calmodulin,** a calcium-binding protein. Undoubtedly, the thigmomorphic response mechanism is similar to other calmodulin/growth responses of plant cells. Ethylene, important in growth regulation, also is important here.

Solar tracking

Some plants move their leaves and flowers toward the sun and track its movement from east to west during the day. The common sunflower got its name because of this trait. In the morning all of its flowers face the east, at noon they lie horizontally facing the zenith, while in the late afternoon and evening they face west toward the setting sun. The phenomenon is called **heliotropism,** although it is not a true tropic response since it does not involve growth. Turgor pressure changes in parenchyma cells account for the movements.

Circadian Rhythms

Many plants exhibit a rhythmic behavior on about a 24-hour cycle, such as the flowers that open in late afternoon every day. This regular repetition of growth or activity on approximately a 24-hour cycle is called a **circadian rhythm.** All sorts of metabolic processes are circadian, such as cell divisions in root tips, and protein or hormone synthesis. **Sleep movements** of leaves are well-known circadian rhythms, as are the opening and closing of night-blooming or day-blooming flowers.

Circadian rhythms are **endogenous,** meaning they are controlled by an internal timing mechanism called the **biological clock** of the plant. Although circadian rhythms are not triggered by an external stimulus, the environment does set and keep the biological clock in harmony with external changes such as darkness and light. The resetting of the biological clock is called **entrainment.** Entrainment, for example, keeps the circadian periodicity of flowering in sequence with light and dark periods even as the day lengths change seasonally. The biological clock keeps the plant responding appropriately for each season by measuring the changing day lengths. The mechanism by which it does this involves the pigment **phytochrome** (see next section).

Photoperiodism

Photoperiodism is a biological response to a change in the proportions of light and dark in a 24-hour daily cycle. Plants use it to measure the seasons and to coordinate seasonal events such as flowering.

Phytochrome

Plants make such adjustments by utilizing the pigment phytochrome, which exists in two forms: **P_r,** which absorbs red light, and **P_{fr},** which absorbs far-red light. Each can convert to the other when they absorb light. During the day, the two forms convert back and forth (P_r becomes P_{fr}, and vice versa), until they reach an equilibrium of 60:40 P_{fr}: P_r in plant tissues. During the night, P_{fr} slowly converts to P_r or else disintegrates. P_r is stable in the dark.

P_{fr} is the biologically active form, acting as the switch that turns on such plant responses as flowering or seed germination. When the threshold concentration of P_{fr} is attained, the response is stimulated. Thus, it is the length of the night period, not the day period, that determines the response. Short nights (meaning long days) favor activities that require large amounts of P_{fr}; conversely, if the night is long (and the day short), more P_{fr} is converted back to P_r and responses triggered by small amounts of P_{fr} are favored. P_r, synthesized from amino acids, is the inactive form.

Photoperiodic responses

Photoperiodism was first studied in relation to flowering. Plants can be described in relation to their photoperiod responses as **short-day, long-day, day-neutral,** and **intermediate-day** plants. Plants that flower in late summer and fall are short-day; they have a critical period of light exposure of less than about 16 hours. Long-day plants are summer flowering and have a critical period of longer than 9 to 16 hours. Day-neutral plants flower in photoperiods of any length, while intermediate-day plants flower only in periods neither too long nor too short for the particular plant (that amount of time is different for each plant studied to date but not classifiable as either long-day or short-day).

Other photoperiodic responses involving the phytochrome system include seed germination and the early growth of seedlings.

Florigen

Because hormones control so many metabolic activities in plants, flowering has long seemed likely to be under the control of one or more hormones. Early experiments sought to determine which part of a plant is sensitive to the light that initiates flowering. The results suggested the presence of a substance that moved from the leaves to the flower buds. Although the substance was not identified then—nor has it been isolated now—it was named **florigen.** Florigen is **the hypothetical flowering hormone;** it may or may not actually exist. Note that flowering most likely is *not* controlled by a single hormone, but is the result of a combination of internal and external signals and responses.

Dormancy

Shoot dormancy

Rarely do all factors of the environment remain suitable indefinitely for plant growth. In the temperate latitudes, for example, breaks in the growing season occur when seasons change, bringing reduced temperatures and shorter days in the autumn and winter. In the subtropics and tropics, where temperatures and day lengths remain equitable all year, water availability may fluctuate between a wet and a dry season. Plants have developed mechanisms to survive during such adverse periods.

One effective mechanism, used by annual plants, is to produce a photosynthetic and flowering structure rapidly and then sink the resources derived from photosynthesis into seed production and distribution. The plant body is no longer useful and is abandoned after protected embryos are produced. The seeds withstand the changes of the next unfavorable growth period and germinate when environmental stimuli indicate favorable growth conditions. Perennial flowering plants also use the seed mechanism, but some retain their photosynthetic and root structures, merely dropping the most vulnerable parts (leaves) during the unfavorable growth period. When one or more of the plant organs undergoes a period in which the growth processes are slowed down or suspended, that state is termed **dormancy.** The growth is reactivated when environmental stimuli are received that, in effect, inform the plant that conditions are again suitable for growth. The signals to break dormancy are extraordinarily precise. External stimuli combine with internal signals to ensure that renewal of growth will occur at the most favorable time. Many plants have internal growth inhibitors that decay slowly over time, such as ABA. Until the inhibitor has dropped to a certain low level, no growth will take place despite external stimuli; both external and internal signals must be correct.

Seed dormancy

Almost all seeds undergo some period of dormancy—if they did not, they would start to grow in the fruits on the mother plant and defeat their principal purposes: dispersal and survival of the germplasm. The period between the formation of the seed and the time when it will germinate is called the **after-ripening period,** which may be a few days or months depending on the plant.

Seeds of plants native to regions with cold winters almost all require an after-ripening period of cold temperatures before they will germinate. This requirement can be met in horticultural and crop varieties by refrigerating the moist seed for a period of time. This procedure is called **stratification.** Dormancy of seeds with hard seed coats often can be broken artificially by **scarifying** the seed—mechanically thinning the seed coat with a file or nicking it with a knife, allowing water and oxygen to penetrate to the embryo.

Bud dormancy

Woody and herbaceous perennials produce dormant overwintering buds in habitats with cold winters. The buds are miniature shoots with apical meristems, leaf primordia, and axillary buds, the whole enclosed by several modified leaves called **bud scales.** The scales protect the embryonic tissues of the bud from mechanical injury and insulate them. In many climates the scales prevent the formation of ice crystals in the young tissues. Bud scales also restrict gas exchange and prevent desiccation. They often accumulate growth inhibitors as well.

Buds start their growth early in the growing season and by midsummer are completely formed. They then undergo a series of physical and physiological changes in preparation for winter. The process is called **acclimation** and is triggered primarily by the shorter days of late summer. Plants that have acclimated to winter are said to be **cold-hardy.** Dormancy is broken in the spring in tree buds by the lengthening days. The buds are the photoperiod receptors.

Senescence

Senescence is the orderly, age-induced breakdown of cells and their components, leading to the decline and ultimate death of a plant or plant part. The timing of senescence is species-specific and varies among the organs of individual plants. Some species of plants produce short-lived flowers whose petals last for only a few hours before shriveling and dropping off, while the leaves of deciduous plants last through long growing seasons before senescing.

Senescence is a metabolic process; therefore, it requires energy. It is not simply the ending of growth. Leaves, for example, move the products of photosynthesis—and their own structural substances—out of leaf tissue into stem and root tissue during senescence and before their vascular connections are severed at abscission. One of the first materials to degrade is the energy-converting pigment chlorophyll. As the bright green color of chlorophyll fades, the yellow-orange colors of the carotenoids become prominent and combine with the red-blue anthocyanins to produce the vivid colors of autumn in the trees and shrubs of the northern deciduous forest.

The role of hormones in senescence is not clear. Not only the kinds, but the proportions of each are important. Ethylene promotes abscission of leaves, flowers, and fruits, while IAA retards senescence and abscission. When days shorten in autumn, IAA production decreases, and ethylene production increases, hastening changes in the cells of the abscission zone. When the degradation of the cell wall materials is complete, nothing remains to hold the leaf to the stem, and with any slight disturbance the leaf falls. Some evidence indicates that a **senescence factor,** presumably an unknown hormone, exists in some plants (like soybeans), but it has yet to be isolated or synthesized.

CHAPTER 14
CELL DIVISION

Prokaryote Cell Division

The continuity of life depends upon the ability of cells to reproduce. In the prokaryotes, cellular reproduction is by **binary fission**, an asexual division of the contents of a single cell into two new cells of approximately equal size. The process is fast and relatively simple: The circular **bacterial chromosome** replicates, and the two new genomes move toward opposite ends of the cell. A new plasma membrane is added between them, dividing the cytoplasm roughly in two, and the cell splits. Each of the two **daughter cells** formed has a complete set of genes and some materials with which to begin an independent life. During periods of active growth, the new cells acquire and metabolize nutrients, grow, replicate their bacterial chromosome, and reproduce once more. In a favorable environment—bathed in the warm rich nutrients of the small intestine, for example—the bacterial cells can divide every 20 to 30 minutes.

Eukaryote Cell Divison: The Cell Cycle

Division isn't as simple in eukaryotes, where linear chromosomes that are contained within a membrane-bound nucleus have to be apportioned equally between two daughter cells. If something goes wrong and they aren't distributed equally, chances are the daughter cells will die for lack of instructions on how properly to conduct the business of life. The eukaryote cell is also filled with organelles and other cytoplasmic materials that must be divided. Small wonder, then, that the process not only is more highly orchestrated, but that it takes much longer to accomplish.

The essentially continuous process of cellular division in body (**somatic**) cells has three significant steps: 1.) the actual division of the nucleus, called **mitosis;** 2.) the division of the cytoplasmic material—**cytokinesis**—into two daughter cells after the nuclear division; and 3.) the **interphase** just before and after division. The division to produce sex cells (**gametes**) is called **meiosis** and involves still other complications, which are discussed later in this chapter.

Stages

The entire sequence of repeating events from one mitotic division to the next is referred to as the **cell cycle.** The cycle has two principal parts: 1.) **interphase**, divided into $\mathbf{G_1}$ (Gap 1), **S** (DNA synthesis), and $\mathbf{G_2}$ (Gap 2) and, 2.) the **M phase**—the combination of mitosis and cytokinesis. Figure 14-1 diagrams the stages in the cycle, and Table 14-1 lists the events of significance in each. Figure 14-2 diagrams the steps in mitosis.

The time to complete the cell cycle varies among species, the tissues in which the cells occur, and general environmental conditions. Of the nuclear division stages, prophase is the longest, and the separation and movement of the daughter chromatids in anaphase is the shortest. Relative lengths might be: prophase 1 to 2+ hours, metaphase 5 to 15 minutes, anaphase 2 to 10 minutes, and telophase 10 to 30 minutes. A fourth interphase stage—$\mathbf{G_0}$—often is present in plants. It can occur at almost any point and is induced by unfavorable growing conditions, such as the onset of cold winter weather or by drought during the summer. It is a holding stage while the plant is dormant.

Sexual Reproduction: Meiosis

A second type of cell division called **meiosis** takes place in multicellular eukaryotes. This is a **reduction division** in which the daughter cells receive exactly half the number of chromosomes of the mother cells.

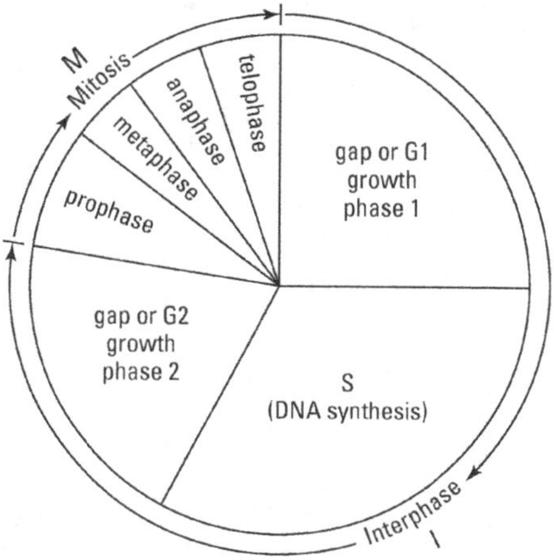

The Cell Cycle

Figure 14-1

Meiosis occurs in the production of **gametes**—the **sperm** of the males and the **eggs** of the females. When a sperm fertilizes an egg, a **zygote** is produced with the appropriate number of chromosomes for the species—in humans (and potatoes) the zygote and the somatic (body) cells produced from it have 46 chromosomes. This is the **diploid** ($2n$) number of chromosomes, half of which have come from the sperm nucleus, half from the egg. The sperm and egg are **haploid** (n); they carry half the number of chromosomes of the body cells (in humans, 23 in each sperm and egg). Meiosis thus makes it possible to maintain a constant number of chromosomes in a species that reproduces sexually by halving the number of chromosomes in the reproductive cells. Meiosis uses many of the same mechanisms as mitosis and is assumed to have been derived from mitosis after the latter procedures were in place in some early organisms millenia ago.

Table 14-1: Steps in the Plant Cell Cycle

Phase	**Process**	**Stage**	**Substage**				
Interphase	G_1 Phase			Cells increase in size; organelles increase in number; enzymes, other substances increase to accompany needs of larger cells; period of biochemical activity			
	✓	**CHECK POINT**	✓	Internal controls on process — proceed or cancel at this point depending upon success of coordination of all steps; signal sent to trigger S phase if all is well			
	S Phase			DNA replicated so cell now has 2 copies of genetic information; proteins (histones) associated with nucleus also duplicated			
	G_2 Phase			Machinery for nuclear and cytoplasmic division assembled; chromatin begins condensation process			
	✓	**CHECK POINT**	✓	Final examination instituted by cellular control system; mitosis proceeds only if all conditions are acceptable and signals sent to initiate process			
				Location of Activities Taking Place In the Cell			
				Chromosomes	Nuclear Envelope	Nucleolus	Other Structures
M Phase	Mitosis	Prophase	Early	Chromatin condenses; 1st threadlike chromosomes appear	Clear zone around it	Distinct	Microtubules scattered around nuclear envelope
			Middle	Shortened; now seen to be composed of 2 separate **chromatids** twisted together and joined at the **centromere** region	Becoming less distinct	Becoming less distinct	Microtubules beginning to align
			Late	Pairs of chromatids lie parallel, joined along their length; centromere region is constricted and two **kinetochores,** one on each chromatid, appear there (these are protein complexes)	Breaks down completely [signals the end of prophase]	Disappears	Microtubules aligned parallel to nuclear surface in *preprophase spindle*

Table 14-1: Steps in the Plant Cell Cycle

				Location of Activities Taking Place in the Cell			
Phase	Process	Stage	Substage	Chromosomes	Nuclear Envelope	Nucleolus	Other Structures
		Anaphase		Chromatids separate into *daughter chromosomes* and begin to move apart toward opposite poles	Absent	Absent	Polar microtubules increase in length, kinetochore microtubules shorten; motor proteins (dynein et al.) using ATP, pull chromatids poleward
		Telophase		Chromosomes elongate, begin to disappear	Membranes derived from vesicles reorganize around each set of daughter chromosomes	Reforms	Spindle apparatus disappears; phragmoplast of microtubules forms on either side of equatorial plane
	Cytokinesis			A cell plate (of vesicles from Golgi apparatus) forms as a disc on the equatorial plane between the two groups of daughter chromosomes; grows outward and joins the cell walls thus forming 2 cells; the cell plate becomes the middle lamella between the 2 new cells when each deposits primary wall materials on it			

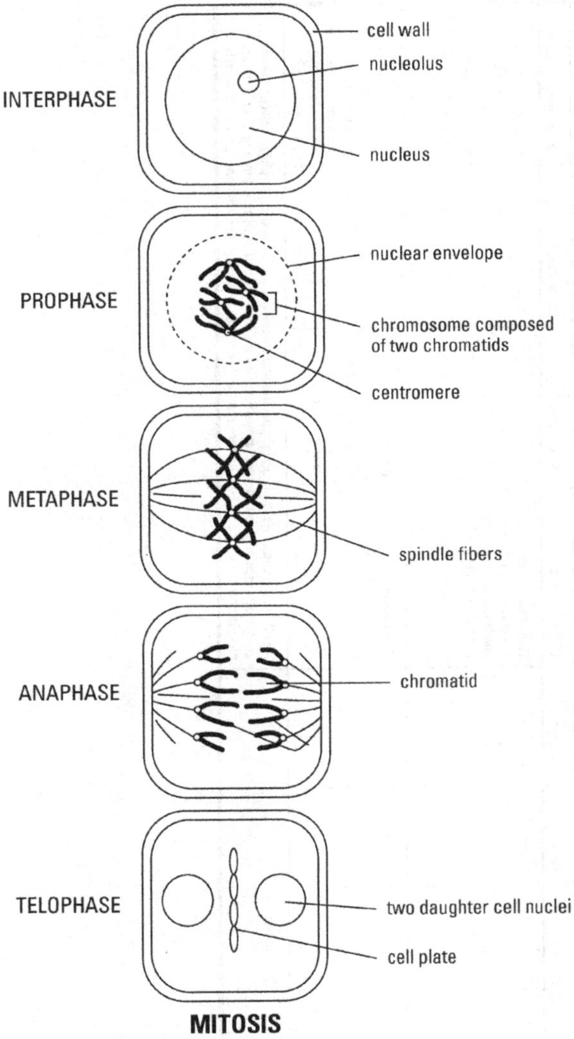
INTERPHASE
cell wall
nucleolus
nucleus
PROPHASE
nuclear envelope
chromosome composed of two chromatids
centromere
METAPHASE
spindle fibers
ANAPHASE
chromatid
TELOPHASE
two daughter cell nuclei
cell plate
MITOSIS

Figure 14-2

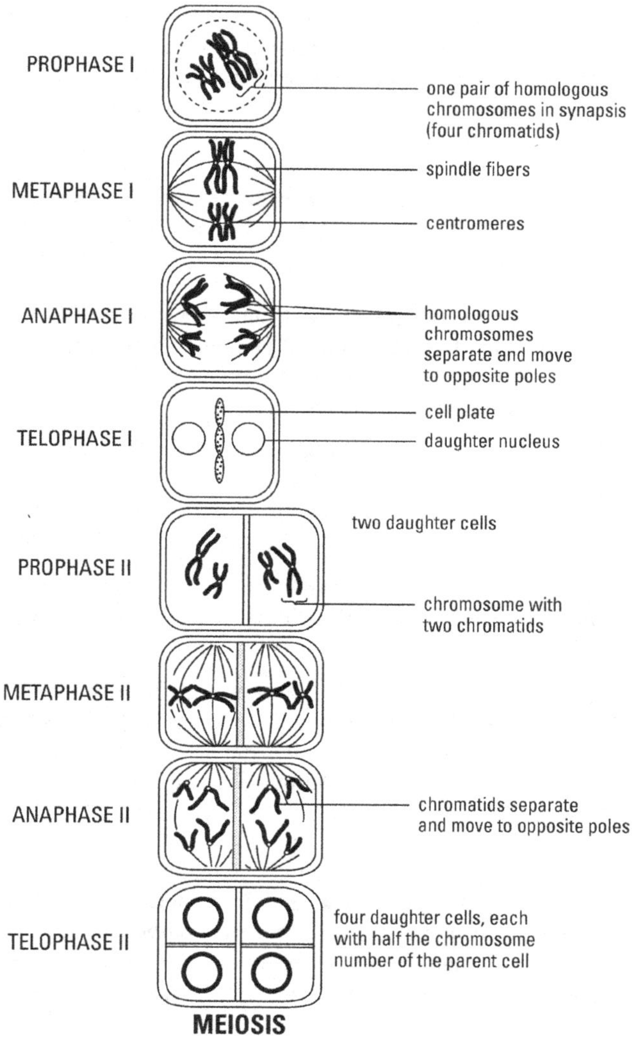
PROPHASE I
one pair of homologous chromosomes in synapsis (four chromatids)
METAPHASE I
spindle fibers
centromeres
ANAPHASE I
homologous chromosomes separate and move to opposite poles
TELOPHASE I
cell plate
daughter nucleus
two daughter cells
PROPHASE II
chromosome with two chromatids
METAPHASE II
ANAPHASE II
chromatids separate and move to opposite poles
TELOPHASE II
four daughter cells, each with half the chromosome number of the parent cell
MEIOSIS

Figure 14-3

Figure 14-3 shows the stages of meiosis. Note that the names for the stages are the same as those of mitosis, with the addition of a numeral to designate either the first or the second divisional stage. *Both divisions are part of meiosis;* not until the final four daughter cells are produced is the process complete.

Meiosis and mitosis have many similarities. There are, however, several fundamental differences. Compare Figure 14-2 (mitosis) with Figure 14-3 (meiosis). In meiosis:

- In Prophase I, **homologous** chromosomes come together in **synapsis** and form pairs called **bivalents** or **tetrads** (because there are four chromatids in the pair); each bivalent has two chromosomes and four tetrads.
- In Metaphase I, bivalents align randomly on the equatorial plane, which means that each daughter cell has an equal chance of getting either the chromosome from the sperm or one from the egg.
- In Anaphase I, the chromosomes separate, each with two chromatids, and move to opposite poles; each of the two daughter cells is now haploid (*n*).
- There is no S phase, and the chromosomes line up immediately in Metaphase II, their chromatids separate in Anaphase II and in Telophase II new cell walls form around the four haploid cells. (Events of the second division are similar to those of mitosis.)

Synapsis in Prophase I is a decisive interval in determining the inheritance of the daughter cells. At this time, **genetic recombination** can occur; that is, daughter cells may receive *combined* traits of their two parents rather than simply the trait from one or the other. This is possible because the phenomenon called **crossing over** often occurs when the chromatids lie together—segments containing similar **alleles** break apart and rejoin to the corresponding segment of the opposite chromatid, thus mixing the traits from individual parents.

CHAPTER 15
GENETICS

Mendelian Genetics

The breeding experiments of the monk **Gregor Mendel** in the mid-1800s laid the groundwork for the science of genetics. He published only two papers in his lifetime and died unheralded in 1884. The significance of his paper published in 1866 on inheritance in peas (which he grew in the monastery garden) apparently went unnoticed for the next 34 years until three separate botanists, who also were theorizing about heredity in plants, independently cited the work in 1900. During the next 30 years, the universality of his findings was confirmed, and breeding programs for better livestock and crop plants—and the science of genetics—were well under way.

At the time of Mendel's work, scientists widely believed that offspring *blended* the characteristics of their parents, but Mendel's painstaking experimentation suggested this was not so. Remember, no one had yet heard of genes, chromosomes, or meiosis, but Mendel concluded from his breeding experiments that particles or "factors" that passed from the parents to the offspring through the gametes were directly responsible for the physical traits he saw first lost in the offspring's generation, then repeated in the next. Closer still to the actual truth, Mendel even hypothesized that *two* factors, probably one from each parent, interacted to produce the results. His "factors" were, of course, the genes, which do, indeed, come in pairs or **alleles** for each trait.

Some say Mendel was lucky, others that his reported results are too good to be true, that he (or someone else) must have fudged the data to make them "come out right." His choice of garden peas was fortuitous. Peas are self-pollinated, and the seven traits he chose to measure are inherited as single factors, so Mendel could establish true-breeding lines for each trait. Thus, he was able to select the parent traits, pollinate the flowers, and count the results in the offspring with no complicating elements. He was mathematically trained, kept accurate records, and applied mathematical analyses (and was among the first to do so with biological materials).

Mendel's first law: Law of Segregation

Mendel did not formulate his conclusions as laws or principles of genetics, but later researchers have done so. Restating and using modern, standardized terminology, this is the information that developed and expanded from his early experiments.

- Inherited traits are encoded in the DNA in segments called **genes,** which are located at particular sites (**loci**, singular **locus**) in the chromosomes. (Genes are Mendel's "factors.")
- Genes occur in pairs called **alleles,** which occupy the same physical positions on homologous chromosomes; both homologous chromosomes and alleles segregate during meiosis, which results in haploid gametes.
- The chromosomes and their alleles for each trait segregate independently, so all possible combinations are present in the gametes.
- The expression of the trait that results in the physical appearance of an organism is called the **phenotype** in contrast to the **genotype,** which is the actual genetic constitution.
- The alleles do not necessarily express themselves equally; one trait can mask the expression of the other. The masking factor is the **dominant** trait, the masked the **recessive**.
- If both alleles for a trait are the same in an individual, the individual is **homozygous** for the trait, and can be either homozygous dominant or homozygous recessive.
- If the alleles are different—that is, one is dominant, the other recessive—the individual is **heterozygous** for the trait. (Animal and plant breeders often use the term "true-breeding" for homozygous individuals.)

Geneticists use a standard shorthand to express traits using letters of the alphabet, upper case for dominant, lower case for recessive. Red color, for example, might be **R** or **r** so a homozygous dominant individual would be **RR**, a homozygous recessive individual, **rr** and a heterozygous individual **Rr.**

Crosses between parents that differ in a single gene pair (such as those that Mendel made) are called **monohybrid crosses** (usually TT and tt). Crosses that involve two traits are called **dihybrid crosses**. Symbols are used to depict the crosses and their offspring. The letter **P** is used for the parental generation and the letter **F** for the filial or offspring generation. $\mathbf{F_1}$ is the first filial generation, $\mathbf{F_2}$ the second, and so forth.

What kinds of crosses did Mendel make to conclude that factors/genes segregate? First of all, he made certain that the plants that he planned to use in the experiment were **pure line** for the trait—that is, that they bred true for the trait for two or more years. (Peas are self-pollinated so he simply grew the plants and examined their offspring.) Other experimenters omitted this step, which confounded their results. Mendel then made a series of monohybrid crosses for each of the seven traits he had identified using parents of opposite traits—tall (TT) vs. dwarf (tt), yellow seed (YY) vs. green (yy) seed, round seed (RR) vs. wrinkled (rr), and so forth. (He, of course, did not symbolize them with letters, but he did know that seeds from his tall pure-line plants would always produce tall plants, seeds from the dwarfs would always produce dwarf plants, and so on.)

P_1 parental generation	TT	vs.	tt
Gametes	T only		t only
F_1 genotype		Tt	
F_1 phenotype		tall	

Mendel then let the F_1 plants self-pollinate: Tt × Tt and in the F_2 generation counted the numbers of individuals with each of the traits. For the tall × dwarf crosses he got 787 tall plants and 277 dwarf plants (6,022 yellow seeds and 2,001 green seeds, and so forth).

P_2 parental generation	Tt	vs.	Tt
Gametes	T and t		T and t
F_2 genotypes		TT, Tt,tt	
F_2 phenotypes		tall, dwarf	
ratio tall:dwarf = 3:1			
actual numbers: 787 tall:277 dwarf; 74%:26% = about 3:1			

An easy way to determine the possible gene combinations is to construct a **Punnett square,** a grid in which all the possible gametes from one parent are listed on one side and those from the second parent across the top. Combine the gametes from the side and the top in the squares, and all of the possible gamete combinations are diagrammed. The previous cross in a Punnett square would look like this:

	T	**t**
T	TT	Tt
t	Tt	tt

You can see from the Punnett square that three of the four gamete combinations will contain at least one dominant allele (T) and that there is only one chance out of four that the recessive (t) can be expressed. Mendel's experimental results fit the phenotypic probability ratio of 3:1. The genotypic ratio, which Mendel didn't know about, is not 3:1, but 1:2:1. That is, 1 homozygous dominant (TT):2 heterozygous dominants (Tt):1 homozygous recessive (tt). The Punnett square shows only the *possible* combinations, not the actual. It provides an easy way to visualize the **probabilities** of a certain combination occurring. In some inherited traits, whether the allele comes from the male or the female parent can make a difference, but in most traits such information does not matter.

After making monohybrid crosses for all the traits and finding that the ratios always approximated 3:1, although the actual numbers of plants and offspring for each cross varied, Mendel concluded that the traits must be carried in pairs that **segregate** (separate) when gametes are formed. This conclusion is now known as **Mendel's first law, the Law of Segregation.**

To confirm his hypothesis, he made another kind of cross, a **backcross,** which mates an offspring with one of its parents. Mendel backcrossed his F_2 tall plants to the dwarf parent and got half tall plants, half dwarf, a 1:1 ratio. If he had backcrossed to the tall parent, what would the ratio have been? Right, *all* tall; that's why breeders today make **test crosses** back to the *homozygous recessive* parent to see if their phenotypically dominant individuals are homozygous or heterozygous.

	t	t
T	Tt	Tt
t	tt	tt

Mendel's second law: Law of Independent Assortment

Mendel also worked with crosses involving two traits—this is where his luck really entered in. The traits he picked are on separate chromosomes (though, of course, he didn't know this). Had they been on the same chromosomes, the ratios he obtained would not have been possible because the traits would always go together in the same gamete unless some cellular tinkering took place.

The mechanisms for figuring out the possible gametes with two traits, filling out the Punnett square, and counting the possibilities are the same—only with more variations possible (see Table 15-1 for potential numbers).

Table 15-1: Possible Two-Trait Genetic Variations

	Monohybrid	Dihybrid	Trihybrid	*n* - hybrid
No. of different kinds of gametes	2	4	8	2^n
Proportion of homozygous recessives in F_2	1/4	1/16	1/64	$(1/2^n)^2$
No. of different phenotypes in F_2	2	4	8	2^n
No. of different genotypes in F_2	3	9	27	3^n

Here's what the cross looks like for two of Mendel's traits combined, flower color and pod characteristics. One allele for each goes in each gamete; purple color (P) is dominant over white (p) flowers, and inflated pods (I) are dominant over constricted (i).

P_1 parental generation	PPII	vs.	ppii
Gametes	PI only		pi only
F_1 genotype		PpIi	
F_1 phenotype		purple inflated	

Self pollinate the F_1 purple flowered, inflated pod plants and what is the F_2 ratio? Not 3:1 anymore. Fill out a Punnett square and see the possibilities. Each gamete gets one allele of each trait, so a dominant purple (P) can have either a dominant inflated pod (I) or a recessive constricted pod (i); ditto the white (p). Thus, four kinds of gametes are possible: PI, Pi, pI, pi and 4 × 4 combinations are possible from the two parents:

	PI	**Pi**	**pI**	**pi**
PI	PPII	PPIi	PpII	PpIi
Pi	PPIi	PPii	PpIi	Ppii
pI	PpII	PpIi	ppII	ppIi
pi	PpIi	Ppii	ppIi	ppii

The phenotypic dihybrid ratio is 9:3:3:1—9 purple inflated, 3 purple constricted, 3 white inflated, and 1 white constricted. (Geneticists now test their results statistically to see if they approach the theoretical 9:3:3:1 and usually use the χ^2 [chi-square] test.)

Mendel drew a conclusion on the basis of his dihybrid crosses that is now known as **Mendel's second law: the Law of Independent Assortment.** It states that during gamete formation the segregation of the alleles of one allelic pair is independent of the segregation of the alleles of other genes.

Mendel confirmed this hypothesis further (as he did in the monohybrid crosses) by backcrossing the F_1 dihybrid to the recessive parent.

The backcross parents	PpIi	vs.	ppii
Gametes	PI, Pi,pI, pi		pi only

The Punnett square for the backcross looks like this:

	PI	**Pi**	**pI**	**pi**
pi	PpIi	Ppii	ppIi	ppii

The phenotypic ratio for the testcross is: 1:1:1:1; that is, 1 purple inflated:1 purple constricted:1 white inflated:1 white constricted—which indicates that the traits have separated and recombined independently of one another.

Intricacies of Inheritance

Continued breeding experiments, better microscopes, and more scientists working in the field have advanced the knowledge of inheritance in organisms and, at the same time, complicated the simple patterns discovered by Mendel. This section covers some of the intricacies.

Linkage and crossing over

Shortly after the genetic community accepted Mendel's Law of Independent Assortment, several exceptions to its operation were found. Most of these exceptions were the result of **linkage** of the genes being studied on one chromosome. When the usual crosses were made (P_1: parents pure-line dominant for two traits × pure-line recessive for two traits), the F_1 individuals of the cross were all dominant and presumably heterozygous. Selfing (transferring pollen from the anthers to the stigma of the same flower) of the F_1 resulted in no predictable ratios, and never the expected 9:3:3:1. Two phenotypes, those of the original P_1 parents, were in high frequency in the F_2 and two other phenotypes, in low frequency, combined the phenotypes of the two original parents. In searching for explanations for the phenomena, the scientists followed the **principle of parsimony**—that is, they looked first for the *simplest* explanation that fits all the facts. In this instance, the simplest interpretation—and the correct one—is that the genes for the traits lie close together on the same chromosome.

Linkage might properly explain the high frequencies of two phenotypes, but what of the low frequency, other combinations? The most logical explanation is that during Prophase I of meiosis when the four chromatids of two homologous chromosomes lie close together, **crossing over** occurs; that is, there is a physical exchange of material between non-sister chromatids and a **genetic recombination** (see Chapter 14). Thus, unexpected gene frequencies occur because genes no longer travel in their previous sequences. The X-shaped location of the crossover is called the **chiasma** (plural, **chiasmata**), and there may be several in each pair.

If genes lie close together on a chromosome, there is less chance of crossing over taking place than if they lie farther apart. After tabulating the frequency of crossing over for known genes, it is possible to construct **linkage maps** of the chromosomes and determine approximate locations of genes.

Incomplete dominance

Incomplete dominance occurs when both alleles in a heterozygous individual are expressed, producing a phenotype different from either single allele. For example, red snapdragons crossed with white snapdragons produce pink snapdragons. A cross of two pinks restores the red and white in a 1 red:2 pink:1 white ratio. The dominant allele does not completely mask the recessive in this case. Although the phenotype is changed, the alleles themselves are unaltered, as can be shown by a backcross in which they segregate and express their original trait in the homozygous condition.

Mutations

A **mutation** is defined as any change in the DNA of an organism—a sufficiently broad definition to include all manner of changes: **deletions** (a piece of the chromosome breaks off and is lost), **translocations** (pieces of material are exchanged between two nonhomologous chromosomes), **inversions** (two breaks occur and the segment in between rotates and reattaches with its gene sequence in opposite direction to the original), **base substitutions** (a different base is substituted for the original), **duplications** (gene sequences are repeated and added to the chromosome), and other changes. **Point mutations (gene mutations)** are changes in DNA that are limited to one base pair; the gene changes and becomes different from its allele. **Chromosome mutations** occur when parts of a chromosome, or whole chromosomes, change.

Polyploidy

A cell or an organism containing more than two sets of chromosomes is called a **polyploid,** which most often forms when homologs do not separate at anaphase I in meiosis (**nondisjunction**). Gametes produced in this fashion will be diploid ($2n$) rather than haploid ($1n$). If two of the diploid gametes unite, the resulting individual will be **tetraploid** ($4n$). Tetraploids are able to reproduce because there is an even number of chromosomes to pair at meiosis—there's simply one set too many. If an odd number (triploid, pentaploid, and so forth) results through only partial disjunction or some other deviation, the individual is usually sterile because the extra set of chromosomes lacks a partner (homolog) during meiotic division. Polyploidy in animals is rare because of this. Because plants commonly reproduce vegetatively, however, polyploidy is common in many plant families (and is especially prevalent in the arctic flora). A particular kind of asexual reproduction termed **apomixis** permits transmission of polyploids through seeds. Apomictic plants form embryos and seeds without fertilization. (Dandelions are apomictic, as are many grass taxa.) Polyploids that form within individuals of the same species are called **autopolyploids.** Those that are produced when two different species cross are **allopolyploids** and **interspecific hybrids.**

Other variations

Numerous other varieties of interactions occur. **Epistasis** (epi = upon), for example, results when the action of one gene masks the expression of a different gene. Some plants have **multiple alleles** for a specific gene, not just the two discussed earlier in this chapter. Others have **polygenic inheritance** in which many genes combine to express a trait. The differences in the trait show a **continuous variation** because none of the genes have a clear dominance over the others. Genes that influence several phenotypic characteristics are termed **pleiotropic** genes.

CHAPTER 16
EVOLUTION

Evolution, as understood by biologists, is the change through time that occurs in **populations** of organisms in response to changing environments. The changes, coded in the molecules of **DNA**, are transmitted from generation to generation and over the history of the Earth have resulted in progressively more complex life forms. The name of **Charles Darwin** and his **theory of natural selection** are inexorably attached to evolution and, together with the mechanisms of genetics, form the basis of the modern theory of evolution.

Darwin's Theory

Simplifying and paraphrasing from Darwin's book, *On the Origin of Species by Means of Natural Selection*, and adding current interpretations, the main points of his theory are: all life came from one or a few kinds of simple organisms; new species arise gradually from pre-existing species; the result of competition among species is extinction of the less fit; gaps in the fossil record account for the lack of transitional forms. These assertions set the stage for the next part of the theory, *why* life evolves: the number of individuals increases at a geometric rate; populations of organisms tend to remain the same size because the resources are limited, and only the fittest survive; the survivors are variable, and those that survive reproduce, perpetuating the favorable traits.

Natural selection, according to Darwin, is similar to **artificial selection.** The environment acted as the selecting force in natural selection. Unlike the relatively rapid selection pressures instituted by breeders, however, natural selection took long periods of time to accomplish change. Darwin was familiar with the then new conclusions of geologists that the Earth was far older than previously thought, which gave his theory of natural selection sufficient time in which to work.

A major problem was an explanation for *how* the favorable selections were perpetuated. In the 1860s the idea that offspring were blends or mixtures of the traits of their parents, the so-called "blending theory of inheritance," was unable to accommodate transmittal of favorable adaptations from one generation to the next. With botanist Gregor Mendel's ideas and the development of genetics, the inheritance portion of Darwin's theory no longer posed a problem.

The Modern Theory of Evolution

The neo-Darwin view of evolution incorporates modern understanding of population genetics, developmental biology, and paleontology, to which is being added knowledge of the molecular sequencing of DNA and the insights it provides concerning the phylogeny of life. The major premises of the **genetic (synthetic) theory of evolution** are: evolution is the change of gene (allele) frequencies in the gene pool of a population over many generations; species (and their gene pools) are isolated from one another, and the gene pool of each species is held together by gene flow; an individual has only a portion of the pool, which came from two different parents, and the portions are different in each individual; the alleles the individual receives are subject to chromosomal or gene mutations and recombinations; natural selection will favor some individuals, who will then contribute a larger portion of their gene combinations to the gene pool of the next generation; changes of allele frequencies come about primarily by natural selection, but migration, gene flow, and chromosomal variations are contributing factors; isolation and restriction of gene flow between subpopulations and their parent populations are necessary for the genetic and phenotypic divergence of the subpopulations.

CHAPTER 17
SYSTEMATICS

Modern Taxonomy Includes Phylogenetics

Systematics is the name for the branch of biology concerned with the study of the kinds of organisms, their relationships to one another, and their evolutionary history. **Taxonomy,** a term often used interchangeably with systematics, is the part of systematics involved in the description, naming, and classification of organisms. **Phylogenetics,** another part of systematics, is the study of the **phylogeny** or evolutionary history of an organism or a group of organisms. Two underlying goals of plant systematics, thus, are to:

- Find, describe, give unique names to, and organize into categories the species of plants of the world (a goal of taxonomy).
- Organize plants and plant groups to reflect their evolutionary relatedness and their descent from a common ancestor (a goal of phylogenetics).

Systematics today is a vigorous and exciting field that has been given great impetus by the discoveries of molecular biologists, who now are describing organisms at their most fundamental level—the DNA sequences of the cells—and providing the systematists new data on which to base their **phylogenetic trees.** Phylogenetic trees are the graphic representation of the evolutionary divergences of organisms that put together on the same branches the organisms most closely related, with oldest ancestors near the base, youngest descendants near the top. The trees obtained from the DNA sequences basically trace the history of how the genes have changed through time. (Figure 17-2 is a phylogenetic tree.)

Naming Plants

Biologists around the world use today a single method with standardized rules to name plants and animals: the **bionomial system of nomenclature.**

The bionomial system of nomenclature

The binomial system in use today gives a single name recognizable throughout the world to each individual kind of organism. The **scientific name** consists of two parts (in Latin): the name of the **genus** (plural: **genera**), plus the name of the particular **species.** The system originated with **Carl Linnaeus** in the middle of the eighteenth century as a shortcut to the cumbersome polynomial system then in use that required 12-word descriptions to be written as part of the name. In the binomial system, the scientific name is *italicized* in print and the genus is capitalized, but the species is not.

Lay people often ridicule scientific names as unpronounceable atrocities, but these same scoffers use many genera names with little complaint. Geranium, chrysanthemum, aster, asparagus, primula, begonia, and rhododendron (as well as hundreds of others) are not only common plant names, but genera names as well. Other common names are recognizable as anglicized versions of such genera names as: *Pinus, Juniperus, Cyperus, Rosa, Hyacinthus, Tulipa, Lilium* and others.

Scientific names are important because they are exact: one kind of plant, one name. Common names vary from place to place and language to language, but scientific names in Latin remain the same and are recognizable anywhere in the world.

Taxonomic hierarchy

The Linnaean **hierarchical system** (see Table 17-1) is a means to group similar organisms together in levels of increasing inclusiveness from the species at the bottom to the most inclusive—kingdom—at the top. Genera are groups of species, families are groups of genera, and so on up the hierarchy. **Taxon** (plural: **taxa**) is a general name given to the members of any level in the hierarchy; in Table 17-1. *Aster* is a taxon or Anthophyta is a taxon or *spectabilis* is a taxon.

Table 17-1: Linnaean Hierarchical System

Kingdom	Plantae
Phylum	Anthophyta
Class	Dicotyledonae
Order	Asterales
Family	Asteraceae
Genus	*Aster*
Species	*spectabilis*
common name	showy aster
scientific name	*Aster spectabilis*

Types of Classifications

Classifications are orderly ways to present information and, depending upon their objectives, can be artificial, natural, or phylogenetic (phyletic), which includes phenetic and cladistic.

Artificial and natural classifications

Classifications that use single or at most only a few characteristics to group plants usually are **artificial** classifications—that is, all the plants in a single group share the same characteristics, but they are not closely related to one another genetically. Popular floras (books to identify plants of a certain area) sometimes group plants using color of their flowers, or their growth form (trees, shrubs, herbs, and so on). Although such books are useful in finding the names of taxa, they provide few clues about relationships among the taxa and hence are not predictive, which means that you can deduce nothing more about the plant other than that it exhibits the characteristics used to classify it. **Natural** classifications group together plants with many of the same characteristics and are highly predictive. That is, by enumerating the characteristics of a plant, one can predict the natural group to which it belongs. Taxonomic floras, for example, identify species, genera, and families by listing as many characteristics as possible concerning anatomy, morphology, cytology, ecology, biochemistry, genetics, and distribution.

Phylogenetic (phyletic) classifications

Phyletic classifications are natural classifications that try to identify the evolutionary history of natural groups. When botanists accepted Darwin's theory of evolution near the end of the last century, the reasons why some groups of plants looked alike became clear: They were related to one another by a common ancestry. The mission of taxonomy since Darwin has become a quest for evolutionary relationships, not just at the lower levels of the hierarchy, but at the upper levels as well.

The evolutionary history of a taxon is called its **phylogeny.** To establish phylogenies, decisions must be made concerning which characteristics are "primitive" and which "advanced"—that is, which taxon is the ancestor of the others. Early phylogenetic classifications were based primarily upon plant morphology and anatomy with great emphasis upon reproductive morphology, which is more stable and less influenced by the environment than is vegetative morphology. Today, taxonomists additionally use the techniques of biochemistry and molecular biology to add details of internal organization and mechanisms to the classifications. But phylogenies, no matter how carefully constructed, are dependent upon someone's interpretation of data, and herein lies the problem: Systematists frequently differ in their interpretations of relationships. A phylogenetic classification is a **hypothesis,** a scientific explanation of the data and, like any hypothesis, is subject to further testing.

Certain assumptions are necessary in phylogenetic classifications. A taxon should be **monophyletic** (all of the members of the taxon should be descendants of a single common ancestor). The **characters** or features used to identify the taxa must be **homologous,** which means that they must have a common origin, but not necessarily a common function. For example, all the parts of a flower—petals, sepals, stamens, and carpels—originate in the same way as leaves from primordia in meristems. Although they now have different functions in the flower (they're not photosynthetic), some sepals and petals structurally resemble leaves. Leaves and the parts of the flower are homologous structures.

Some features that look alike do not have a common origin and are said to be **analogous.** An example of analogous structures is the prickles on two groups of succulent desert plants, the cacti and the euphorbs. Cacti have spines that are modified leaves; euphorbs have thorns that are modified branches. Spines and thorns look alike and are functionally similar in that both keep animals from eating the plants. Spines and thorns are analogous. This example of analogy is also an example of **convergent evolution.** The cactus family and the euphorb family both developed the same morphology in response to a desert environment—the cacti in North and South America, the euphorbs in Africa and Asia. The families are not related and have no recent common ancestor.

Numerical taxonomy (phenetics). Systematists have tried many ways to make phyletic classifications more subjective. When computers became readily accessible in the 1960s, **numerical taxonomy** or **phenetics** became a popular approach. In practice, measurements were made of a large number of characters of a taxon, at least 60 per plant and often 100 or more. No special importance was attributed to any one of the characters. After the measurements were complete on hundreds of individuals, the data were analyzed statistically with computer programs and cluster analysis or other methods to show purported natural groupings of plants with overall similarities. Systematists' interpretations were thought to be minimized in this fashion.

Cladistics. Cladistics is the most popular method of classifying organisms today. In contrast to phenetics, in which similarities are sought using as many characters as possible, cladists look for patterns using **derived** character states (that is, features that have evolved from an ancestral character group). The intent is to find groups of organisms that share a common ancestor and to diagram the relationship of the groups, called **clades,** in a **cladogram** (see Figure 17-1). The branching points (nodes) separate groups that have diverged in the evolutionary past from a common ancestor. All the taxa below the node lack the character state, all those above it retain it. Homologous (inherited) characters are chosen to categorize an organism and its character states. The states are hypothesized to be either ancestral or derived (evolved), and the cladogram is a test of the hypothesis.

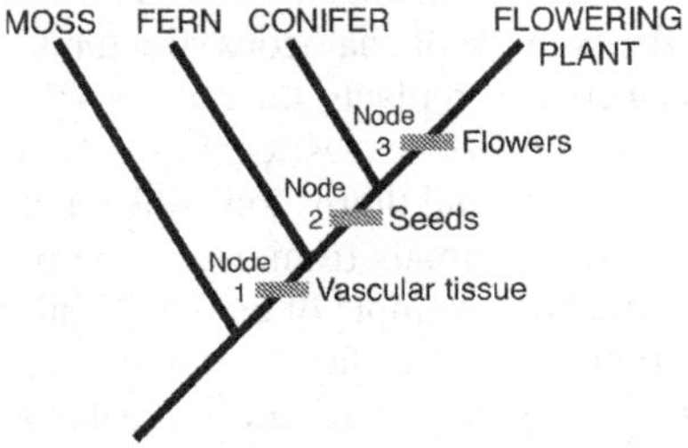

Figure 17-1

Molecular biology and phylogeny. The most promising developments in formulating a phylogeny for the entire tree of life come today from molecular biology, where new tools and techniques allow researchers to use as character states the **molecular sequences** of **amino acids** as well as those of **nucleotides** in **nucleic acids.** The latter is the most fundamental of comparisons for, of course, the genes control the structure of life itself. The closer the similarity in sequences of molecules among groups of organisms, the closer the relationship of the groups. Widely different sequences indicate a different evolutionary history and ancestry.

Some assumptions made by the users of molecular sequencing include:

- Phenotypic (outward appearance) evolutionary changes accompanied by genetic (hereditary) changes occur over time in organisms.
- Long time intervals result in the accumulation of more changes.
- Organisms that have the most similarities in their gene sequences are more closely related than those with fewer; they have had a shorter time in which to evolve different phenotypes and genotypes.
- The groups with widely different sequences must have separated at an earlier time in the evolutionary past.

Plants Among the Diversity of Organisms

Classification schemes are in a state of flux because of the availability of large volumes of data generated by molecular sequencing of DNA and RNA. As might be expected, disagreements among biologists are common. For example, not all biologists believe widely different-appearing and behaving organisms should be grouped together just because they have similar DNA base pair sequences. But, the cladists do (and are willing to debate the doubters).

Major groups and current ways of grouping of organisms

In the middle of the eighteenth century, Linnaeus' ideas transformed biological classification. In the latter half of the nineteenth century, Darwin revolutionized biology with an irrefutable theory of evolution. At the end of the twentieth century, molecular sequencing is changing the phylogeny of the entire tree of life. Appropriately enough, a major adjustment has already been made at the roots of the tree: There appear to be *three* main lines of development from the primitive milieu. The hodge-podge of prokaryotes (unicellular, non-nucleated organisms) clearly belong to two separate groups: the **Bacteria** and the **Archaea (Archaebacteria).** The nucleated organisms (eukaryotes)— plants, animals, and so forth— fit in a separate lineage, the **Eukarya** (Figure 17-2). The Linnaean hierarchy is modified and a new name added for these three "super kingdoms," the **Domain.**

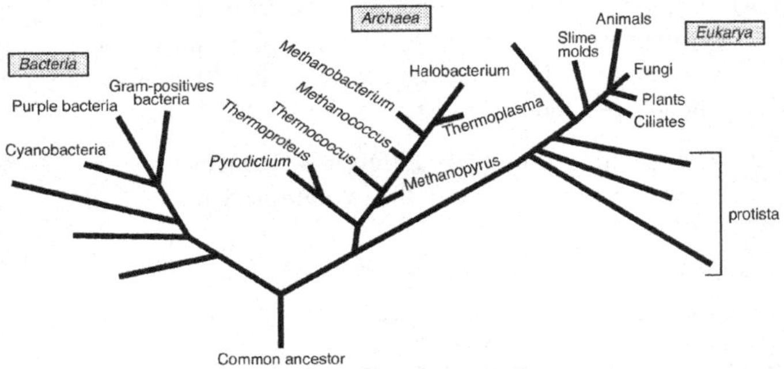

Figure 17-2

Controversial as this change has been, shifts among groupings of the Eukarya are even more controversial, not because the data are suspect, but because biologists differ on how best to organize the new information with the old. Organisms in the five-kingdom approach of the recent past are now distributed among four kingdoms of the *Domain Eukarya* and the two domains of prokaryotes, *Domain Bacteria* and *Domain Archaea* (see Table 2-2 and Table 2-3 in Chapter 2).

This change among groupings brings up a problem for botanists. What do you do if the organisms you study are evicted from the plant kingdom? Are you still a plant scientist if you no longer study plants? Many of the ousted groups are included in plant biology textbooks with the justification that, because the groups share many of the features of plants, it's appropriate for botanists to study them.

Classifications are based on current knowledge, which is constantly changing, so rearrangements are bound to occur along with differences of opinion about what belongs where. Rarely do all parties agree. Tables 2-2 and 2-3 (in Chapter 2) outline how classifications of plants have changed over time. Some of the old group names survive the advent of new classification schemes and are useful ways to discuss informally some groups. Figure 1-1 (in Chapter 1) summarizes the way a majority of botanists believe the plants are related to one another and to the rest of the organisms.

Features Used in Classifying Groups of Organisms

Biologists use the following features of organisms to identify the major groupings of current classifications. (This book does not discuss animals and animal-like protists beyond placing them in general perspective.)

- Presence or absence of a defined nucleus
- Unicellular or multicellular with specialized organelles
- Mode of nutrition
- Presence or absence of a cell wall
- Composition of the cell wall
- Motility
- Mode of reproduction
- Kind of life cycle

Nucleus

The most basic division of organisms separates the living world into two groups on the basis of those possessing and those lacking a defined **nucleus** (plural: **nuclei**). The nucleus is an organelle, which contains the major portion of the genetic material (DNA) of the cell and is surrounded by a **nuclear membrane.** The genetic material of

Prokaryotes is not contained within a membrane-bounded nucleus. Eukaryotes all have nuclei.

Cellularity

The form (morphology) of an organism can be **unicellular** (one-celled) or **multicellular** (many-celled). Some unicellular organisms form filaments (strings of cells), others form sheets of cells held together by pectins, and still others form colonies that give a superficial resemblance to multicellularity. Unicellular organisms do not form **tissues** (similar cells organized into a functional unit) nor **organs** (groups of tissues organized for a particular function). Some organisms alternate a unicellular stage with a multicellular stage in their life cycles. Eukaryotic organisms have **organelles,** membrane-bounded structures within their cells specialized to perform certain functions.

Nutrition

All organisms need a source of **energy** to fuel their **metabolism,** the chemical processes that maintain life. Organisms obtain their nutrients for metabolism in one of two basic ways: 1.) **Autotrophs** are able to make the organic compounds they use for metabolism directly from inorganic materials; and 2.) **Heterotrophs** are unable to do this and obtain their nutrients from the organic materials manufactured by autotrophs. Some autotrophs are **photoautotrophs.** They use radiant energy from the sun in the process of **photosynthesis** to manufacture organic compounds. **Chemoautotrophs** use chemical energy in **chemosynthesis,** oxidizing inorganic compounds to manufacture organic nutrients. **Chloroplasts** are present in the photoautotrophs, absent in the chemoautotrophs. Animals are heterotrophs; they **ingest** (swallow) their food and then digest it internally. Fungi are heterotrophs, which release digestive enzymes into their surroundings and then **absorb** the nutrients into their cells. Many protists use **phagotrophy,** a type of nutrition in which single cells ingest food particles. Some fungi (and other organisms) are **saprophages,** heterotrophs that break down the organic materials of dead organisms.

Cell wall

Animals and the animal-like protists have no cell walls, but most other organisms (with a few exceptions) have some kind of wall made from a variety of materials. Almost all of the prokaryote cells have walls, and a major distinction between the Bacteria and the Archaea is the presence of **peptidoglycans (glycoprotein polymers)** in the Bacteria and their absence in the Archaea cell walls. Fungi cell walls are made of **chitin,** the substance that makes the exoskeletons of lobsters, crabs, cockroaches, and other arthropods hard. The basic material of plant cells (and those of many algae) is **cellulose. Lignin, suberin, waxes,** and many other substances may be deposited additionally.

Motility

Plants in general and some animals don't move around; they are **sessile** (attached) to a substrate. But, many plant and sessile animal cells are **motile,** and they move using a variety of techniques. There are motile organisms in all of the kingdoms, so motility per se does not distinguish groups, but the kind and location of the devices employed for movement do determine groups. The organelle that propels most cells is the **flagellum** (plural: **flagella**) or, in the terminology of some biologists, the **undulipodium** (plural: **undulipodia**). A smaller, shorter flagellum is a **cilium** (plural: **cilia**). The flagella are long threads of protoplasm that extend outside of the cell and have the capability for limited movement. The prokaryotes have a single-fiber flagellum that rotates; the flagella of eukaryotes are bundles that consist of nine pairs of **microtubules** wrapped around a central pair (a 9 + 2 configuration). A sliding action moves the microtubules.

Type of reproduction

Reproduction is the creation of new individuals from existing ones and can be either **asexual**—without special sex cells (**gametes**)—or **sexual,** in which gametes fuse to produce new individuals. Gametes are usually **haploid** (with a single set of chromosomes) and their fusion (**fertilization**) results in a **diploid** (with two sets of chromosomes) **zygote** (the cell formed by the fusion of two gametes).

Variations of both sexual and asexual reproduction are legion throughout the living world. Asexual reproduction occurs in some members of all the kingdoms, whereas sexual reproduction is present in all but the Archaea. Many types of asexual reproduction exist. **Fission,** a splitting in two of the cell, is one type of asexual reproduction. In prokaryotes, division of the genetic material accompanies fission, whereas it does not accompany fission in the eukaryotes. Yeasts and some other organisms **bud,** simply by pushing out and breaking off pieces of the cell. **Spore-formation** is a widespread method of asexual reproduction in which single-celled **spores,** formed in specialized structures called **sporangia,** are produced in large numbers. They may undergo a resting stage first, or produce new individuals directly. Sexual spores are produced in some organisms. (See Figure 17-3.)

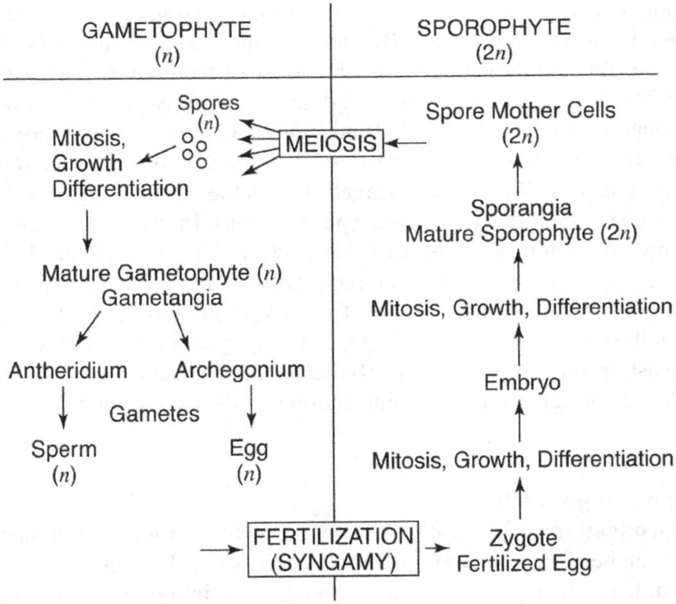

Figure 17-3

Life cycle

Three basic types of life cycles differentiate major groups of organisms (see Figure 17-4). All are variations on a general theme in which haploid cells alternate with diploid in the stages of the life cycle. Thus, meiotic (reduction) cell divisions alternate with fertilization (fusion of gametes). The three life cycles are:

- **Zygotic meiosis:** The individual organisms are haploid, and only the zygote is diploid. The zygote produced by fertilization immediately undergoes meiosis, producing more haploid individuals. This life cycle appears in all fungi and some algae.
- **Gametic meiosis:** The mature, common individuals are diploid and produce haploid gametes that fuse. The zygote divides by ordinary mitosis, producing the adult diploid individuals. Animals, some brown and green algae, and many other organisms maintain this type of life cycle.
- **Sporic meiosis:** Also called **alternation of generations** because during the life cycle two kinds of individuals switch or alternate as the common individual, one diploid, one haploid. (See Figure 17-3.) In plants the diploid individual, called the **sporophyte,** produces **spore mother cells** that divide by meiosis producing haploid **spores.** The spores germinate and produce haploid gametophytes. The latter then produce the haploid gametes, which fuse in fertilization, forming the diploid zygote that matures into the adult sporophyte. In addition to plants, this form of life cycle is present in many algae.

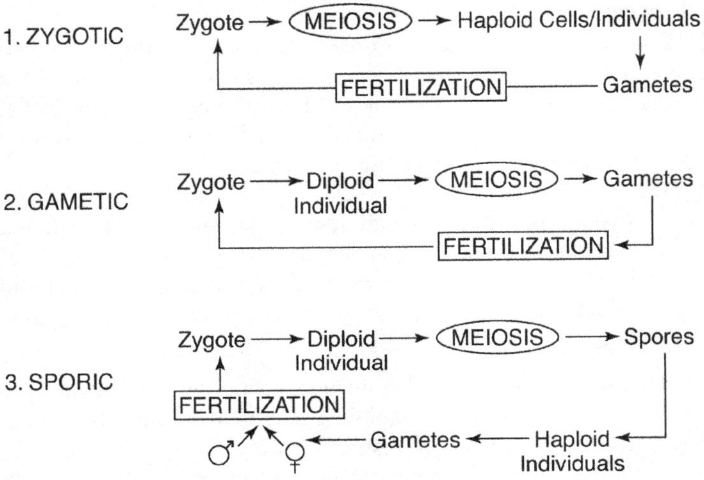
1. ZYGOTIC
Zygote
MEIOSIS
Haploid Cells/Individuals
FERTILIZATION
Gametes
2. GAMETIC
Zygote
Diploid Individual
MEIOSIS
Gametes
FERTILIZATION
3. SPORIC
Zygote
Diploid Individual
MEIOSIS
Spores
FERTILIZATION
♂
♀
Gametes
Haploid Individuals

Figure 17-4

CHAPTER 18
PROKARYOTES AND VIRUSES

General Characteristics of Prokaryotes

The prokaryotes are the most abundant organisms on Earth, and their biomass undoubtedly outweighs all the rest of the organisms together. Although they are too small to be seen individually without powerful magnification, they and the results of their activities are everywhere; without them life on Earth would cease. They have persisted for 3.5 billion years exploiting every possible inorganic and organic habitat—the first 2 billion years alone with no other kinds of organisms. In so doing, they have evolved ways to make a living in each. They manage by being metabolically diverse, morphologically small, cellularily simple, and genetically versatile. They are the dispersers and the recyclers of the Earth's materials and great parts of the human economy depend upon either finding ways to make use of the prokaryotes or ways to get rid of them. Table 18-1 summarizes the basic features that separate the three domains of life.

The division of the prokaryotes into two domains poses many problems, not the least of which is the inclusiveness of the name "bacteria." Technically, "bacteria" aren't *all* of the old "bacteria;" when used appropriately today the fascinating extremophiles are excluded by the term. Some microbiologists suggest the use of "Eubacteria" (eu = true) as a domain and common name to distinguish one specific group, but the practice is not universally accepted. Some clarification may be necessary, therefore, in using—and interpreting others' use—of the word "bacteria."

Table 18-1: Distinguishing Features of the Three Domains of Life

Feature	*Bacteria*	*Archaea*	*Eukarya*
size	10 — 100 µm	1 — 10 µm	cells 10 — 100 µm
tissues	absent	absent	mostly present
nucleus	absent	absent	present
nuclear envelope	absent	absent	present
DNA	— single "bacterial chromosome" (genophore) —		linear chromosomes contain DNA + proteins
number of RNA polymerases	one	several	several
introns	absent	some genes	present
membrane-bound organelles	absent	absent	present
mitochondria	absent	absent	present
major lipids with phytanol side chains	ester-linked	ether-linked	ester-linked
site of photosynthesis	— chromatophores on cell membranes —		in chloroplasts
peptidoglycan in cell wall	present	absent	absent
metabolize methane	no	yes	no
number of RNA polymerases	one	several	several
flagella	— bacterial flagella of flagellin protein; rotate —		complex flagella of tublin + other proteins; termed undulipodia
asexual cell division	— direct; fission or budding; no microtubules —		mitosis; with microtubules
sexual reproduction	— "sexual" recombination; transfer from donor to recipient —		two partners; an alternation of meiosis & fertilization resulting in haploid and diploid nuclei
oxygen requirements	strict anaerobes, anaerobes & aerobes		mostly aerobes
metabolism	— most diverse of all organisms — (see text)		all use the same basic oxidation pathways — Krebs cycle, electron transport chains

Structure

Until light microscopes with better lenses and electron microscopes with higher magnifying capabilities were developed, microbiologists knew little about the structure and considerably more about the chemistry of the organisms they studied. A typical bacterial cell is illustrated in Figure 18-1 with the major features named. The cell is most of the features found in eukaryote cells.

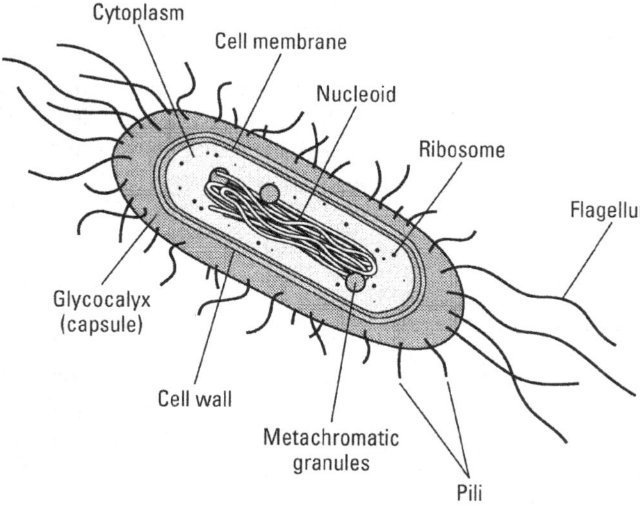

Figure 18-1

When growth conditions become unfavorable—when nutrients become scarce, for example, or the environment dries—many bacteria produce **endospores** adding thick walls around the circular DNA together with a bit of cytoplasm. The spores resist high temperatures, desiccation, chemical disinfectants, ultraviolet radiation, X-rays, boiling for several hours and are the reason bacteria sometimes survive even in "sterile" hospital environments.

Bacteria are basically unicellular with simple shapes: short rods or **bacilli** (singular: **bacillus**), spheres or **cocci,** or spiral, elongated cells, **spirilla.** The single cells often are linked together into ribbon-like **filaments,** or bead-like **chains** of cells; some taxa form flat, sheet-like **colonies,** others produce stalked, branching ones.

All except the mycoplasmas have a **cell wall** composed of disaccharides and peptides (amino acids) together with a unique compound not found in eukaryotes: **peptidoglycan.** The latter substance is present in the Domain Bacteria and absent in the Domain Archaea, making it a good diagnostic feature. The **gram stain,** a dye that reacts with peptidoglycan and proteins of the cell walls, effectively divides the Domain Bacteria individuals into two major groups, *gram-positive* and *gram-negative* members. (In the days of light microscopes when not much beyond general rod-coccus-spirilla shape was discernable, bacteriologists struggled to find good diagnostic features; the chemistry of the walls proved to be one.)

Reproduction

The principal mode of reproduction is an asexual separation of one bacterium into two. There are, in addition, several mechanisms that make possible the exchange of genetic material, the DNA, among and between bacterial cells. None, however, are as elaborate as the mitosis–meiosis choreography of gene exchange in the eukaryotes.

Asexual

Three common types of asexual reproduction are present:

- **Binary fission** (the most common): The "chromosome" replicates and the two genomes move to opposite ends of the cell. The old cell walls then grow inward between the two, pinching the cell apart—no mitosis, no microtubules. The whole process is over within 30 minutes to three hours.

- **Fragmentation:** This occurs when filaments of cells break into separate pieces or *fragments*.
- **Budding:** An outgrowth (*bud*) pushes out from the cell, pinches off, and then enlarges into a new cell.

Gene exchange

With no nuclei, there can be no sexual reproduction in the prokaryotes, but there is an exchange of DNA. In one type, **conjugation,** *conjugation ("sex") pili* (bridges of cytoplasm) form between cells, and some DNA is passed from the donor to the recipient cell.

In bacterial communities some DNA exists outside of cells, presumably left when the cells die, or more probably, excreted into the environment by living cells. This "free" DNA is picked up bacterial cells in another kind of gene exchange, **transformation.**

A third type of exchange—**transduction**—results when **bacteriophages** (special kinds of **viruses**) bring into bacterial cells the DNA from their previous viral host. (Researchers in *biotechnology* use the same method to introduce new genes into host organisms.)

Mutation

Random changes in the DNA are common. These **mutations** of the genetic code alter the response of the individual to its environment. If the mutation is deleterious, the individual dies, but favorable mutations spread rapidly as the cells divide repeatedly and often.

Prokaryote Metabolism

The prokaryotes are the most metabolically diverse of all organisms and have some exotic ways to obtain and channel their needs. Organisms need carbon for building cells and energy to fuel the process; eukaryotes, in general, all follow the same basic metabolic pathways whereas prokaryotes use a variety of materials and pathways—some employed by no other organisms. The terms for energy and carbon acquisition are not standardized among biologists and microbiologists, and to make matters worse, are inconsistently used concerning the separation of energy and carbon acquisition. Table 18-2 lists some basic terms frequently used by plant biologists.

Table 18-2: Metabolic Requirements and Terms

Requirement and Source	Name
Energy source	
sun	photosynthesizers (photoautotrophs)
chemical compounds	chemotrophs
Carbon source	
carbon dioxide (CO_2)	autotrophs
from organic compounds	heterotrophs
Electron donors in respiration	
inorganic molecules	lithotrophs
organic molecules	organotrophs
Nutrition	
inorganic materials	autotrophs
organic materials made by other organisms	heterotrophs
	saprophytes (saprobes, saprotrophs, saprophages) symbiotrophs mutualism commensalism parasitism
Oxygen (O_2) Source	
require free O_2	aerobes
live in the absence of free O_2	anaerobes
	facultative
	obligate (strict)

Photosynthesis

Cyanobacteria. The blue-green algae, or cyanobacteria, are an important group of prokaryote photosynthesizers. They use the same pigment, chlorophyll *a*, as the green plants and have similar carotenoids, but in addition have accessory pigments called **phycobilins.** One of these is blue, **phycocyanin** (the source of the group name), the other a dark red, **phycoerythrin.** (One group of cyanobacteria is common in northern and alpine areas where in spring they grow on the melting snow banks, turning them a deep pink color.)

Notes
plants, algae, cyanobacteria
most organisms get energy this way; most are chemoheterotrophic
majority of the prokaryotes; most are saprobes
hydrogen atoms the electron donors
animals, fungi, prokaryotes, some protista; also carbon from the same molecules
cause decay and recycling of essential elements in the ecosystem
use materials from dead organisms
two organism live together in close association in a symbiosis both partners benefit one benefits with no effect to the other one (the parasite) benefits with some harm to the other (the host)
photosynthesis traps energy; respiration releases it
oxygen the final electron acceptor in respiration
use O_2 when present; if no O_2, switch to anaerobic respiration
have only the anaerobic pathway; most die if O_2 present

The chlorophylls are contained on thylakoid-like membranes with water the hydrogen donor, oxygen the by-product—as in plant photosynthesis.

The cyanobacteria are common in the marine **plankton** community (huge numbers of tiny aquatic organisms that float near the surface), and their presence in the early prokaryote world undoubtedly was a factor in changing the atmosphere from oxygen-free to oxygen-rich, paving the way for arrival of eukaryote oxygen users.

Purple and green bacteria. These bacteria are anaerobes, some of which use sulfur compounds as the electron donors (purple sulfur and green sulfur bacteria) and others of which use such organic compounds as alcohols and fatty acids (purple nonsulfur and green nonsulfur bacteria). Photosynthetic pigments include carotenoids and several bacteriochlorophylls, but no phycobilins. The green sulfur bacteria are *photolithoautotrophs* because their energy comes from the sun (photo), their carbon from CO_2 in the air (auto), their electrons from inorganic sulfur compounds or H_2 gas (litho). Photosynthesis in the purple bacteria appears to be ancestral to Photosystem II of plants, that in the green bacteria a forerunner of Photosystem I.

Nitrogen cycle

The prokaryotes play a central role in the nitrogen cycle, not only in getting nitrogen *into* the cycle, but getting it *out* as well. The actions of the prokaryotes drive the cycle. They mediate the following processes.

Nitrogen fixation. Prokaryotes are the only organisms capable of fixing atmospheric N_2. That is, they form ammonia (NH_3) from gaseous N_2. The bacteria use some of the fixed nitrogen, and the remainder goes into the soil ecosystem for use by all. When the nitrogen-fixers die, their nitrogen-containing compounds are recycled by other organisms. Some nitrogen-fixers are chemoautotrophic soil bacteria, both anaerobes and aerobes, that either live free or else live symbiotically within the root nodules of legumes or in nodules of a few other plants. Cyanobacteria also live in lakes and ponds, in the oceans, and in numerous mutualistic relationships with fungi and plants, in lichen bodies, for example, or with cycads and ferns.

Nitrification. Nitrification is the two-step conversion of NH_3 to nitrite (NO_2^-) and then to nitrate (NO_3^-). The energy released in the process is used by the chemolithoautotrophs to reduce CO_2. Plants are able to assimilate nitrate, but nitrites are toxic to them.

Ammonification. *Nitrogen mineralization* or **ammonification** is the name of the process that the organisms of decay—chiefly saprophytic bacteria and fungi—use to decompose nitrogen-containing organic molecules, which release nitrogen as ammonia in the process. (Mineralization is the term for the conversion of organically bound nutrients into plant-available, inorganic forms, while **decomposition** is a more general term for the breakdown of organic matter.) Without mineralization the world would soon run out of raw materials for organisms, and life would cease. The nitrogen cycle turns because decay organisms exist.

Denitrification. Denitrification is the opposite of nitrogen fixing and nitrification; denitrifying bacteria return nitrogen to the atmosphere as N_2O (nitrous oxide) or gaseous nitrogen (N_2). Anaerobic bacteria carry out denitrification when oxygen diffuses too slowly through the soil to meet the microbial respiration demand. Nitrate replaces oxygen as the needed electron acceptor in these instances.

Extreme environments

The *Archaea* are divided into three large groups based on their physiological diversity. The **methanogens** are strict anaerobes that produce methane (CH_4), **halophiles** are chemoorganotrophs that *require* 12–23 percent salt to grow, and one group of **thermophiles** grows best at temperatures over 80° C , the other group at 0°C and lower. *Pyrolobus* grows at 113° C, the hottest known temperature to maintain an organism; *Thermoplasma* species combine low acidity and high temperature and live at pH 1–2 and temperatures of 60° C.

The methanogens are the only organisms that produce methane. They use ammonium (NH_4^+) as a nitrogen source and get their carbon from CO_2. The supplies of natural gas used today come from the metabolism of methanogens of the past, and all of the methane in the air today (around 1.6 ppm) comes from the metabolism of current methanogens.

The *Archaea* are more closely related to the eukaryotes than to the *Bacteria* and probably evolved later than the *Bacteria*. Our understanding of how many there are and where they live is limited. By virtue of their metabolic versatility, many live in remarkable environments—hot springs, hydrothermal vents at the bottom of the ocean, in the gut of cows, in vats of sulfuric acid, in sewage disposal tanks. Recent discoveries in soils of *r*RNA fragments with molecular sequences similar to those in the *Archaea* suggest that the *Archaea* may be more common than we now know.

Systematics

The oldest known fossils are suspected to be bacteria, ancestors of the modern prokaryotes.

Fossil record

Prokaryotes were the first organisms when the atmosphere of the new Earth was **anoxic** (without oxygen) and there was extensive volcanic activity. Bacterium-like filaments have been found in 3.5 billion-year-old rocks in western Australia, and bacterium-like spheroids of the same age occur in South Africa. Fossil **stromatolites** are present in many ancient sedimentary rocks worldwide. Stromatolites are layered columns or mushroom-shaped domes a few inches wide and a foot or more in height that formed when bacterial mats (composed primarily of cyanobacteria growing in primeval ponds) trapped sediments. Together sediment and bacteria solidified into rock over the ensuing eons. Stromatolites are forming today in many places in the same way by descendants of the same organisms.

Phylogeny

Data from molecular sequencing of DNA has changed—and continues to change—the ideas concerning bacterial relationships. At this time there is no clear consensus among microbiologists concerning the lineages among the prokaryotes. With only an estimated 10 percent of the bacteria named, and the majority of those identified not yet studied in detail, the task appears formidable. Stay tuned.

Ecology

Attributing life-or-death importance to organisms too small to be seen without great magnification is difficult, but consider that the prokaryotes:

- Decompose complex organic molecules and return to the soil and air the elements needed for growth of all organisms.
- Participate in complex biogeochemical webs that concentrate minerals—iron, manganese, copper, and others.
- Maintain soil fertility by fixing atmospheric nitrogen, thus assuring the supply of available nitrogen for protein and nucleic acid synthesis by all organisms.
- Are the base of food webs on land and in the oceans.
- Are crucial links in the sulfur, phosphorus, carbon, oxygen, and nitrogen cycles.
- Using novel metabolic pathways, both discharge into the atmosphere and extract from it *all* of the major reactive gases: nitrogen, oxygen, carbon dioxide, carbon monoxide, sulfur-containing gases, hydrogen, methane, and ammonia.

Human Interest/Economics

Humans have a love–hate relationship with the prokaryotes. They cause terrible diseases, but by recycling the indispensable materials that all organisms need, make life possible on Earth.

Plant pathology

Among the other accomplishments of the prokaryotes is their ability to parasitize all manner of plants and animals. Human bacterial diseases include tuberculosis, Lyme disease, black plague, cholera, botulism, pneumonia, and hundreds of others. Bacteria cause particularly vicious disease in plants of all sorts. The *wilt* diseases are produced by bacteria that live within the xylem vessels, plugging them and preventing water from reaching the upper part of the plant, which causes wilting. After filling the vessel openings, the bacteria next attack the cell walls, rupturing the cells, causing the plant to collapse. *Crown-gall*, a cancerous tumor on stems, is caused by bacteria that inject DNA plasmids into the host cell nuclei, thereby taking control over hormone synthesis—and acquiring a home in the process.

Within the number of pathogens are the **mycoplasmas**—bacteria that lack cell walls. They are the smallest organisms able to live independently. They are 0.2–0.3 μm in diameter, live in the sugar solutions carried in the phloem of vascular plants, cause over 200 extremely destructive diseases of plants, and were completely unknown for decades because they passed through standard laboratory filters and were invisible in ordinary light microscopes. Researchers grew old and gray trying to determine what was happening in their supposedly cell-free solutions before the existence of mycoplasmas was known.

Genetic engineering

Genetic engineers use **recombinant DNA** techniques learned from manipulating bacterial genomes to cut and splice packets of genes. **Bacteriophages,** viruses that infect bacteria, often are used to carry genes between organisms. **Plasmids** with genes of interest can be isolated and the genes duplicated by the billions in industrial applications to produce vaccines and hormones or, in agricultural applications, to develop plants and animals with desirable characteristics. Opinions vary concerning the ethics of doing so. Proponents minimize the risks to the environment, but the potential to wreak havoc remains a disturbing possibility.

The prokaryotes are a mish-mash of contradictions. Some are the smallest of living organisms, but they outnumber all others on Earth. They are the most abundant organisms, but can't be seen with the naked eye, and even light microscopes reveal only gross details; they are structurally the simplest of organisms, but they alone live and thrive in the most extreme environments imaginable. They have been on Earth the longest of all organisms, but we have only begun to identify the kinds extant around us. They cause the deadliest diseases, and the most efficacious antibiotic drugs. We probably could survive without cheese and yogurt or pickles and sauerkraut, which are possible because bacteria produce lactic acid and vinegar, but life itself could not survive without the bacteria.

Viruses

Although often studied by plant biologists, viruses are *not* living organisms because they:

- Are not cellular and have no cytoplasm, membranes nor organelles.
- Can't metabolize; they lack the enzymes necessary for protein synthesis and energy transfer.

- Don't increase in size (they don't grow).
- Don't respond to external stimuli.
- Aren't motile.

But, they *are* able to reproduce within a host cell by using the metabolic equipment of the host. Outside of a cell they are simply a collection of inert molecules of nucleic acids and proteins (which can be denatured easily). They are extremely small, between 20 nm–400 nm, the size of large molecules.

Viruses are **obligate intracellular parasites** that live within the cells of all kinds of organisms, frequently to the detriment and ultimate death of the host cells. Their presence may trigger a response within the host that produces disease symptoms. Each virus can enter only the cells of hosts with receptors specific for it; humans don't get canine distemper, and dogs don't get polio—both virus-caused diseases—because the proteins don't match. Viruses attack hosts whose genomes are most like their own.

Bacteriophages (or simply **phages**) are viruses that invade bacteria. They have very small genomes consisting of only a few genes. A still smaller replicating particle, the **viroid,** is a small bit of RNA without a protein coat. Viroids are about a tenth the size of small viruses and replicate like viruses using the mechanisms of the host cell. Viroids have been identified as causing some infectious plant diseases and probably are responsible for many animal diseases as well.

Plant diseases

The viruses that invade plants do so by entering an open wound or other breaks in the surface or from the actions of an animal invader. The first virus to be isolated and described was the tobacco mosaic virus, which earned its investigator a Nobel prize in 1946. Over 1,000 plant diseases are attributed to viruses and viroids and more than 400 kinds of viruses are involved. Some plants harbor several kinds of viruses and show no effects while others, such as Rembrandt tulips, owe their unique flower colors to the presence of a virus in their cells.

The symptoms of viral diseases usually are systemic rather than localized because the virus spreads throughout the plant in the phloem and from cell to cell through the plasmodesmata. There is no evidence that viruses can penetrate cell walls. In plants infected by some kinds of viruses, only the actively dividing cells of the growing points seem to be virus-free, presumably because the meristem cells divide faster than the virus can move through the plasmodesmata into tip cells.

The universal plant response to a viral infection is reduction in size—stunting of the whole plant, small leaves, and decreased flower and seed production—due to upset of normal metabolism and interference with translocation of materials in the phloem. Symptoms include yellowing of leaves and mottling, leaf spots, wilting, and tumors resulting in abnormal flowers and leaves.

Virus-infected plants are weakened and more susceptible to other diseases. Except for tobacco mosaic disease, there is no evidence, however, that viruses are spread from plant to plant by direct contact of plant parts. Most are spread by insects, nematodes, or other animals (such as slugs and snails) or by soil-inhabiting fungi. Humans contribute to the distribution by making and growing cuttings from diseased plants.

There are no vaccines with which to inoculate plants against viral diseases. Control, at present, consists of stopping the spread (such as burning diseased crop plants and potential weed hosts, sterilizing tools used to make cuttings, destroying the seeds of infected plants, and controlling insect and nematode vectors by insecticides and soil fumigation).

CHAPTER 19
FUNGI: NOT PLANTS

Overview: A Kingdom Separate from Plants

The **fungi** (singular, **fungus**) once were considered to be plants because they grow out of the soil and have rigid cell walls. Now they are placed independently in their own kingdom of equal rank with the animals and plants and, in fact, are more closely related to animals than to plants. Like the animals, they have **chitin** in their cell walls and store reserve food as **glycogen.** (Chitin is the polysaccharide that gives hardness to the external skeletons of lobsters and insects.) They lack chlorophyll and are heterotrophic. Familiar representatives include the edible mushrooms, molds, mildews, yeasts, and the plant pathogens, smuts and rusts.

Most fungi are terrestrial, multicellular eukaryotes, the body (**soma**) of which is a mass of thread-like filaments called **hyphae** (singular, **hypha**), which collectively form a **mycelium** (plural, mycelia). When the fungus reproduces, specialized hyphae pack together tightly and form distinctive **fruiting bodies,** or **sporocarps,** from which sexual spores are released. The ordinary edible mushrooms are the fruiting bodies of fungi. Fruiting bodies are temporary structures in the life cycle; the primary body of all fungi is in reality the diffuse, wide-spreading mycelium.

The fungi reproduce by spores, both asexual and sexual, and the details and structures of the sexual process separate the kingdom into four phyla (see Table 19-1). The **zygote** is the only diploid phase in the life cycle; meiosis occurs shortly after the zygote is formed—hence the life cycle is an instance of **zygotic meiosis.** Chemical signals, **pheromones,** are exchanged among fungi, especially between pairs preparatory to sexual reproduction.

Fungi are **heterotrophs,** which release digestive enzymes into their surroundings and **absorb** nutrients back. Some fungi are **saprobes (saprophytes)**, as important in decomposition as the bacteria; others are **symbiotrophs,** living in symbiotic association with plants, animals, protists, and cyanobacteria. Well-known symbioses are: **lichens** that are associations of fungi and green algae or cyanobacteria; **mycorrhizae,** associations of fungi and plant roots; and **endophytes,** fungi and plant leaves and stems. Some fungi are **parasites (fungal pathogens)** and responsible for diseases of both plants and animals. Complex life cycles involving one or more hosts have developed between fungal pathogens and their hosts.

The Earth's largest living organism may be a fungus: either the mycelium reported from Washington state that covers 1,500 acres (but probably is disjointed and broken) or the one in Michigan that covers 37 acres (and is estimated to weigh 110 tons—the weight of a blue whale).

Table 19-1: The Four Phyla of the Fungi Kingdom

Phylum	*Common Names*	*Familiar Species*	*Number of Species*
Chytridiomycota	chytrids, water molds	*Allomyces*	1,000
Zygomycota	bread molds, zygomycetes	*Mucor, Rhizopus, Pilobolus*	2,000
Ascomycota	sac fungi, truffles, morels, blue-green molds, powdery mildew, chestnut blight	*Penicillium, Saccharomyces, Morchella, Claviceps, Aspergillus*	32,000
Basidiomycota	basidiomycetes, mushrooms, rusts, smuts, puffballs, bracket fungi	*Agaricus, Puccinia, Ustilago, Polyporus, Boletus, Amanita*	22,300

Characteristics

All fungi have some features in common, but other special structural and reproductive features separate the four phyla (see Table 19-2).

Structure

The fungi are **eukaryotic** and have membrane-bound cellular organelles and nuclei. They have no plastids of any kind (and no chlorophyll).

The hyphae of the fungi are of two general kinds: Some are **septate,** and are divided by **septa** (walls) that separate the cylindrical hypha into cells; in the **nonseptate** fungi, the hypha is one long tube. (The septa are perforated, however, permitting the cytoplasm to flow throughout the length of the filament.) Mitosis occurs in the nonseptate hyphae, but there is no accompanying **cytokinesis** (division of the cytoplasm) so the hyphae are **multinucleate** (with many nuclei). The special name for this condition—an organism or part of an organism with many nuclei not separated by walls or membranes—is **coenocytic,** and the organism is a coenocyte.

A few fungi—called by the general name yeasts—are single-celled, and nonfilamentous much of the time. The only flagellated cells in the kingdom are the flagellated gametes of the chytrids.

Metabolism

The fungi are all **heterotrophic**, but unlike animals and many other heterotrophs that *ingest* their nutrients as bits or bites of food, the fungi secrete digestive enzymes into their surroundings, in effect digesting their food outside of their bodies. They then can *absorb* the smaller particles and incorporate the nutrients into their own cells. Some are **parasites** obtaining nutrients from living organisms, but more are **saprobes (saprotrophs)** that digest and recycle materials from dead organisms.

Table 19-2: Characteristics of the Fungi Phlya

Phylum	*Habitat*	*Flagellated Cells*	*Plant Diseases/ Pathogens*	*Walls*	*Chitin*	*Hyphae*	*Asexual Reproduction*	*Specialized Cell Where Nuclear Fusion Occurs*	*Sexual Spore*
Chytridiomycota	water	yes	black wart of potato, brown spot of corn	lacking in some	yes	aseptate, coenocytic	zoospores	None occurs	none
Zygomycota	mostly terrestrial	no	soft rot of many taxa	yes	yes	aseptate, coenocytic	nonmotile spores	fusion of two gametangia	zygospore in zygosporangium
Ascomycota	mostly terrestrial	no	Dutch elm disease, chestnut blight	yes	a few with cellulose	septate	budding, conidia, fragmentation	ascus	eight ascospores
Basidiomycota	mostly terrestrial	no	black stem rust of wheat, white pine blister rust	yes	yes	septate	budding, conidia, fragmentation	basidium	four basidiospores

In addition to potent digestive enzymes, some fungi manufacture powerful alkaloids that, when ingested by humans, assail the nervous system, causing hallucinations and even death. The "death angel", *Amanita*, is one such well-known poisonous fungus; ergot (*Claviceps purpurea*) is another.

Fungal hyphae, like the roots of vascular plants, grow primarily at the tip, elongating and branching repeatedly. The filaments are in direct contact with their environment, obviating in the fungal body the need for separate absorbing and conducting systems (and precluding the need for storage tissues). Materials readily pass through the plasma membrane and cell walls of the hyphae along their entire length, although the most active metabolism and material exchange is concentrated near the hyphal tips. Most of the cytoplasm is located at the tips also.

Reproduction

Nonmotile sexual and asexual **spores**—microscopic in size—are the common means of reproduction and the primary agents of fungal dispersal. They are readily carried in air or attached to the bodies of insects and other animals and are not resistant structures like bacterial endospores. Although they can withstand desiccation, they are killed by heat. Sexual spores often require a period of dormancy after they are formed, but asexual spores usually germinate and produce new hyphae whenever and wherever moisture is available.

Asexual spores are produced in special hyphae called **sporangia** in the zygomycetes and **conidia** in the ascomycetes and basidiomycetes. Unlike many organisms that produce embryos, the fungal spores form hyphae directly with no immature or embryonic stage between spore and adult.

Sexual reproduction

Among fungi, there are no female and male individuals, and no eggs and sperm. Physiological differences among the hyphae do exist, however, and result in different mating types; only compatible strains fuse. In the zygomycetes the strains are designated simply as (+) and (–). Haploid (n) gametes are produced by mitotic division from haploid (n) parent nuclei in specialized hyphae called **gametangia.**

In the ascomycetes and basidiomycetes, sexual reproduction starts with hyphae from two mating strains fusing, but the nuclei remain independent within the merged cytoplasm. The name for this process is **plasmogamy,** and the cells with the two genetically distinct haploid nuclei are called **dikaryons.** In genetic shorthand dikaryotic cells are $n + n$ rather than the $2n$ of diploid cells. At some point, the nuclei combine, mixing the DNA from the two separate mating types. This type of fertilization is called **karyogamy,** the union of nuclei following plasmogamy.

Mycologists frequently use the term **syngamy** for the process of **fertilization;** both syngamy and fertilization, however, mean the same thing: the union of two haploid gametes to form a diploid **zygote.** In most fungi, karyogamy is followed almost immediately by a reduction division (meiosis) that restores the haploid chromosome number to the resultant spores and the new hyphae that are produced when the spores germinate.

Nuclear division

Both meiosis and mitosis in the fungi are different from nuclear division in most other organisms. Before the nuclear material is divided in plants and vertebrates, the nuclear membranes (nuclear envelopes) disintegrate, and the DNA condenses into discrete chromosomes that divide and move into new cells. In the fungi, the nuclear membranes do not disintegrate, and in many taxa no discrete chromosomes appear nor do centrioles develop. In some fungi, spindles form outside of the nucleus and move into it, while in others a spindle apparatus forms within the nucleus. Lacking the elaborate apparatus of more advanced organisms, the nuclei of fungi simply elongate, constrict near the middle, and pinch or tear apart into two daughter nuclei.

Life Cycle

The predominant phase in the life cycle of fungi is haploid, the zygote is the only diploid cell in the entire cycle. This is called a **zygotic life cycle** and is the type prevalent in algae and some protists, in addition to the fungi.

Systematics

Fungi are separated into phyla on the basis of their reproductive structures. Because some fungi have never been observed to reproduce sexually, they have no place in the classification. Until their sexual reproduction is identified they are placed in Fungi Imperfecti (Deuteromycetes). DNA sequencing is giving the mycologists answers. Most of the taxa so far sequenced and classified are ascomycetes, with only a few basidiomycetes and fewer still zygomycetes. Another problem group is the Chytridiomycota (chytrids), which arguably may belong with the Protoctists, not the fungi.

Phylogeny

No clearcut ancestral lineage for the fungi has been established, but on the basis of molecular DNA sequencing and morphological evidence it seems likely that the fungal life style arose many times from different protoctists. The fungi and animals are on the same originating branch. Within the modern fungi, the chytrids are the oldest of the group with the ascomycetes and basidiomyctes closely related and on a different, more recent line from the zygomycetes.

Fossil record

Evidence of fungi growing within the cells of 400-million-year-old Silurian-Age vascular plants suggests an early origin for the fungi. The first fungi developing from very early eukaryotes undoubtedly were unicellular; coenocytic filamentous forms were a later development.

An interesting proposal postulates that a symbiosis between early fungi and early plants permitted the plants to establish themselves on land before they had evolved roots with which to absorb vital water and minerals from the soil. The fungi could do this for them and already were associated with some plants, hence the start of the mycorrhizal association.

Ecology

Wherever there is moisture, moderate temperatures, and a supply of organic food there are fungi. Since they digest their food outside of their bodies, they literally live within their food supplies. When the area around them is depleted, they grow into a new supply. They occur worldwide, although there may be more taxa in the tropics—an assertion that is difficult to support because while there are an estimated 1.5 million species of fungi, less than 10 percent of them have been described. About 500 species are marine; the rest are terrestrial with several thousand described symbionts and plant and animal pathogens.

Fungi usually are the primary decomposers in their natural habitats and are capable of digesting a wide array of organic materials—including, unfortunately, some substances of economic importance to humans. Most are saprobes, but some, like their animal relatives, attack living prey, a notorious example being the fungus that sets hyphal traps, ensnares and then digests nematodes. Many fungi are parasitic and the major pathogens of many crop plants such as corn and wheat.

Symbiotic Relationships

Two important symbioses involve fungi: the mycorrhizae that occur on the roots of almost all vascular plants and the lichens that have evolved entirely different body forms from those of their symbionts.

Mycorrhizae

Fungi and the roots of almost all vascular plants form mutualistic associations called **mycorrhizae** (singular, mycorrhiza). The fungus gets its energy from the plant, and the plant acquires an efficient nutrient absorbing mechanism—the actively growing hyphae that penetrate regions of the soil untapped by root hairs. Phosphate uptake especially is increased when mycorrhizae are present.

Two general types of mycorrhizae occur, differentiated by whether the hyphae live *within* the cortical cells of the roots or remain *outside* the cells: **endomycorrhizae** (endo = within; myco = fungus; rhizae = roots) and **ectomycorrhizae** (ecto = outside). Zygomycete taxa are components of most endomycorrhizae while basidiomycetes and a few ascomycetes form ectomycorrhizae.

Lichens

The symbiotic relationship of fungi with either algae or cyanobacteria produces a body—a **lichen**—so distinctly different from either of its symbionts that it is treated as a separate organism. The fungal hyphae give the lichen **thallus** its characteristic shape, and the cells of its photosynthetic partner are dispersed among them. While the algal or cyanobacterial member can live independently, the fungus cannot, so the fungus in essence is a parasite on the photosynthesizer in the lichen thallus. The fungus, however, provides a "home" for the photosynthetic cells as well as absorbing water and nutrients that the photobiont uses. This makes the symbiosis mutualistic as much as parasitic in the view of some biologists.

Life is becoming precarious for lichens in many urban environments today. Many lichens are intolerant of air pollutants. They have no means of getting rid of the elements, toxic or useful, which they absorb. Sulfur is particularly toxic to many, and sulfur dioxide released from burning coal has eliminated many susceptible species from cities. Lichens can be used as biomonitors—and warnings—of the quality of the air we breathe.

Plant Pathogens

Many of the fungi are pathogens that infect plants and animals causing diseases of many kinds. The life cycles of many of these are complex and involve two or more host plants.

Rusts

The **rusts** are specialized basidiomycetes that are parasites on plants. They have complex life cycles, and some produce five different kinds of spores in addition to basidiospores. Many rusts are **heteroecious** and complete their life cycles on two different kinds of host plants whereas **autoecious** parasites produce all of their different kinds of spores on a single host species. Well-known examples of heteroecious rusts are wheat rust, white pine blister rust, and cedar-apple rust.

Smuts

Smuts are parasitic basidiomycetes that produce powdery masses of black spores enclosed in a membrane. This membrane is often found in the ovaries of species of grasses, or on their leaves. The smut life cycle is less complex than that of the rusts, and only one other kind of spore in addition to basidiospores is produced. All smuts complete their life cycle on only one kind of plant. Smuts live as saprobes in the soil, however, and readily attack developing seedlings planted in infected soil. Corn smut is common in the Midwest. Despite the unappetizing appearance of the spore masses and their dust-like texture, membrane-enclosed corn smut spore masses are considered delicacies in some cultures and are eaten either boiled or fried.

Yeasts

Yeasts are unicellular fungi that reproduce asexually by budding, a process by which a new cell is formed from a bulge or "bud" that enlarges and pinches off from the parent cell. The nuclear material is divided by mitosis, and the new cell receives a nucleus and cellular organelles before severance from the parent. Yeasts are found in all three of the fungal phyla, but most are ascomycetes. Many are filamentous most of the time, and change to the yeast growth form only occasionally.

Yeasts are of great importance to the baking and brewing industries with particular strains guarded and nurtured closely, because the products of the yeast metabolism give the distinctive flavors to the brews and cause the bread dough to rise in a predictable fashion.

CHAPTER 20
PROTISTA

A Mixture of Life Forms

The Kingdom Protista (Protoctista) is a hodgepodge of organisms with little relationship to one another. Most members of this kingdom have features of one or more of the other kingdoms, but not enough to place them legitimately into any one of these as currently defined. Cladists suggest rearranging the tree of life to accommodate the protista and redefining its branches based on gene sequencing and other new data. Three new kingdoms—each equal in rank to the present animals, fungi, and plants—would be added and the plant and animal kingdoms expanded to include the rest of the protista. The proposal is still in the pro versus con heated debate stage. This could be simply another esoteric controversy among biologists if not for the fact that the organisms under discussion are some of the most important in the world. Scores of protists are the single-celled, photosynthetic, primary producers of marine and freshwater food chains—and have been for the past billion years. In addition, all of the protists provide basic clues to how life evolved on the early Earth; the kingdom is a collection of experiments on how to manage energy and conduct life. Within the protista are found the basic plans that were adopted, adapted, and modified into the life we see around us today.

With classifications in such a state of flux, which organisms one includes in the Kingdom Protista, and how to group them, are matters of choice. Table 20-1 follows the lead of several popular textbooks. Animal-like protists have been omitted from the table. They include such organisms as **Sarcodina:** amoebas, forams, radiolarians; **Flagellates:** dinoflagellates, euglenoids, zooflagellates (trypanosomes, *Giardia)*; parasitic **Sporozoans;** and **Ciliates** like *Paramecium.* Those protists of particular interest to plant biologists usually include the fungi-like and plant-like organisms. As always, *follow your instructor's views on the systematics of the protista.*

Table 20-1: Major Phyla of Protists and Some of Their Characteristics

Group	*Phylum*	*Organisms*	*Chlorophylls*	*Locomotion*	*Reserve Carbohydrate*	*Cell Wall Composition*	*Habitat*
FUNGUS-LIKE							
Water molds	Oomycota	oomycetes, *Saprolegnia, Phytophthora, Plasmopara, Pythium*	none	2 flagella in zoospores & male gametes only	glycogen	cellulose cell walls	marine, freshwater & terrestrial; plant pathogens
Slime molds							
plasmodial	Myxomycota	myxomycete	none	2 flagella, in gametes; amoeboid	glycogen	none on plasmodium	terrestrial
cellular	Dictyosteliomycota (Acrasiomycota)	dictyostelids	none	amoeboid	glycogen	cellulose	terrestrial
PLANT-LIKE (ALGAE)							
Cryptomonads	Cryptophyta	cryptomonads, cryptophytes	none, or chlorophylls *a* and *c*; phycobilins; carotenoids	flagella, unequal, subapical	starch	no cell wall; protein plates	marine and freshwater; cold water

Red algae	Rhodophyta	red algae, coralline algae	chlorophyll *a*; phycobilins; carotenoids	none	floridean starch	cellulose embedded in gelatinous matrix	mostly marine, warm waters;
Haptophytes	Haptophyta	haptophytes, coccoliths	chlorophylls *a* and *c*; carotenoids, especially fucoxanthin	none or 2 flagella		scales (coccoliths) of organic material;	mostly marine; cause toxic "blooms"
Diatoms	Bacillario-phyta	diatoms	none or chlorophylls *a* and *c*; carotenoids mainly fucoxanthin	none or 2 flagella; apical	chrysola-minarin	silica	marine and fresh-water
Chrysophytes	Chrysophyta	chrysophytes, yellow-green algae	none or chlorophylls *a* and *c*; carotenoids mainly fucoxanthin	none or 2 flagella; apical	chrysola-minarin	none or silica scales; some cellulose scales	mainly fresh-water, a few marine

(continued)

Table 20-1: Major Phyla of Protists and Some of Their Characteristics *(continued)*

Group	*Phylum*	*Organisms*	*Chlorophylls*	*Locomotion*	*Reserve Carbohydrate*	*Cell Wall Composition*	*Habitat*
Brown algae	Phaeophyta	brown algae, kelps, rock-weed, *Fucus*	chlorophylls *a* and *c*; carotenoids mainly fucoxanthin	2 flagella only in reproductive cells	laminarin transported as mannitol	cellulose in algin; some algin; some with plasmo-desmata	almost all marine, in cold waters
Green algae	Chlorophyta	green algae	chlorophylls *a* and *b*; carotenoids	none, 2 or more flagella	starch	proteins, noncellulose carbo-hydrates, cellulose; some with plasmo-desmata	marine and fresh-water many symbionts

*The characteristics as tabulated apply to the majority of organisms in each group; with such a disparate collection of organisms, however, there are numerous exceptions to the general rule.

Significant Features of the Phyla: Fungi-Like Protista

Slime molds

The Kingdom Protista was established in the 1860s as a place for the slime molds that are plant-like in forming spores in multinucleate, erect, sporangia and having cellulose in their cell walls, animal-like in having an amoeboid stage in their life cycle during which they creep about ingesting their food, and fungal-like in general appearance and habits.

Like fungi these slime molds grow in damp, organic-rich sites. Rotting logs or decaying plants on the forest floor are favorite habitats. Their amoeboid form is frequently a brightly colored orange or yellow blob of viscous, slippery protoplasm that streams slowly into a network of branching, anastomosing projections that move the whole mass forward. This is the feeding stage and bacteria, yeasts, fungi, or bits of vegetation are incorporated into the mass as it moves. Two principal groups of slime molds are recognized, with a third unrelated group closely associated:

- **Dictyosteliomycota** form a motile mass of protoplasm—a "slug"—by aggregating individual amoeboid cells that retain their identity in the slug, hence their common name, **cellular slime molds**.
- **Myxomycota,** the **plasmodial slime molds,** lose their cell membranes when they come together and the nuclei float freely in the combined, membrane bound mass of cytoplasm, which is called a **plasmodium.** (The slug of the Dictyosteliomycota is a false plasmodium or a **pseudoplasmodium.**)
- **Labyrinthulomycota,** the **slime nets,** are basically unicellular, but live together in colonies. They secrete a membrane outside of their cells, which forms a network of filaments through which the cells travel. The net, sometimes several centimeters in diameter, resembles the plasmodium of the slime molds, but if examined microscopically the cells in their tracks can be seen as distinctively different.

The three groups are neither related to the fungi nor to each other, but they have in common heterotrophy and a sporangia form in their life cycle.

The myxomycetes are reproductively more advanced than the cellular slime molds and form a true plasmodium that can be several centimeters in diameter and look to the untrained eye like a patch of yellow vomitus spewed over a decaying log on the forest floor. The plasmodium is multinucleate and diploid. When food supplies are limited or the environment dries, the plasmodia cease their streaming and may form thick-walled structures called **sclerotia** (singular, sclerotium) in which the protoplasm can withstand adverse environmental conditions. Growth is resumed with the return of more favorable circumstances.

Both asexual and sexual reproduction occurs, but the initiation clues to each process remain obscure. Under some conditions, the plasmodium produces erect sporangiophores with sporangia on their tips. Meiosis takes place and haploid spores result. Like the sclerotia, the spores are resistant and able to sustain the slime mold over adverse growth periods. When conditions again are suitable for growth, the spores germinate. Some develop into amoebas that move about, feeding. Others become flagellated gametes. After a period of time, the cytoplasm of a pair of genetically different amoeboid or flagellated cells fuse, but individuality of the nuclei is retained, a process known as **plasmogamy. Karyogamy,** the fusion of the nuclei, soon follows after both types of unions; the resultant cells are the **zygotes.** Growth of a zygote and the repeated division of its nuclei by mitosis result in the characteristic multinucleate, diploid plasmodium.

Oomycetes

Another distinctive group, the **oomycetes,** includes the water molds and some other taxa. Some are simple unicellular forms, but there are, as well, branched, coenocytic, filamentous fungi-appearing individuals in the group. They have cellulose in their cell walls, reproduce sexually by oogamy, and asexually with biflagellate zoospores.

A large number of the 700 species are aquatic, both marine and freshwater, hence the common name "water molds" for the group. Even the terrestrial members form flagellated zoospores when water is available in their habitat. Among the terrestrial forms are several highly destructive plant pathogens—downy mildew of grapes and late blight of potatoes are caused by oomycetes. Less well known, but as significant economically, are several oomycetes that live in the soil and attack, and ultimately kill, the roots of many commercially grown fruits. Large areas in southern California, for example, are unsuitable for avocado growth because of oomycetes in the soil.

Phylogeny of the fungi-like protists

The slime molds represent three stages in development towards multicellularity: a single, coenocytic mass of protoplasm; a mass of protoplasm in which separate cells float; and a third mass of protoplasm in which the cells are enclosed and move in membranous structures. All of these are able to come together and function in a mass that resembles a multicellular organism.

Significant Features of the Phyla: Algae

Some marine algae—the kelps—look like plants with stem-like stalks supporting "leaf" *blades* and a *holdfast* that anchors the kelp to rocks on the sea bottom. Growing offshore in colder northern waters they sometimes attain a height of 50–60 meters—the largest protists. Most of the algae are constructed more simply and are smaller—many are unicellular and microscopic in size. The algae, like the rest of the protists, have evolved a wondrous array of body forms and lifestyles—unicellular, motile free-living; complex aggregations of cells in colonies; filaments, nets, coenocytic tubes; microscopic, slimy green threads—all are common algal forms.

The variable body structure is not useful taxonomically to separate the major groups, nor are the reproductive structures particularly helpful. Instead, the following five features are more definitive and used in algal classification:

- The photosynthetic pigments they contain.
- Kind of stored foods.
- Materials composing the wall.
- Number, kind, and position of the flagella.
- Details of cell structure, such as the shape of the chloroplasts or the presence of pyrenoids or eye spots.

Today, a sixth characteristic is added:

- The molecular sequences of their DNA and RNA.

Data obtained by using this newest approach are shaking the algal phylogenetic tree and revising many long-held ideas of relationships within the algae. Newly acquired data, however, have not changed the belief that members of the green algae orders Coleochaetales and Charales of the Class Charophyceae are the closest living relatives of the plants.

The green algae are green because they contain chlorophylls *a* and *b* and carotenoids in the same proportions as the green plants. Like the green plants, they store starch inside the plastids and most have cellulose cell walls. Some have reproductive cells that are structured with two whiplash flagella—like green plant sperm cells. These common features led biologists in the past to the green algae as the protistan ancestors of green plants—if one defines "plants" as "land plants" (embryophytes). With today's available data, some biologists say that algae *are* green plants of the aquatic kind and suggest a general category of Chlorobiota to include *all* green plants, aquatic and terrestrial. There are strong advocates for both approaches, but no clear majority for one over the other.

CHAPTER 21
BRYOPHYTES—THE NON-VASCULAR PLANTS

Bryophytes Are Land Plants Without Vascular Tissues

Bryophytes are small, low-growing plants of mostly moist environments in the temperate and tropical zones where they grow on the ground and as epiphytes on the trees and undergrowth. In the alpine and boreal zones, bryophytes often are the dominant life form. Some bryophytes are desert dwellers, and a few are aquatic; none are marine.

The bryophytes are of botanical interest because their ancestors apparently were among the first land plants. The existing species today have some green algal features and some vascular plant attributes making them intermediates—more complex than green algae, but not quite vascular plants. Once thought to be monophyletic, the bryophytes are now recognized as having three independent lineages.

Characteristics

Bryophytes are *plants* because they are photosynthetic with chlorophylls *a* and *b*, store starch, are multicellular, develop from embryos, have sporic meiosis—an alternation of generations—and cellulose cell walls. Some mosses have simple water and food conduction-type cells (but these are not the same as the xylem and phloem tissues of vascular plants). They have no lignified cell walls (like wood) for strength, so the plants remain small. Neither do they have leaves, stems, or roots. They absorb water from their surfaces by capillarity. The "leaves" of leafy liverworts and mosses are undifferentiated tissues and lack stomata, and the moss "stems" lack vascular tissues.

Within the bryophytes, the three lineages show a progressive development of body structure—the liverworts have the most simple, the hornworts intermediate, and the mosses the most specialized thalli.

The gametophyte is the dominant, independent, long-lived generation (the leafy moss plant and the flat, green thalli of the liverworts and hornworts are the gametophytes). The sporophyte is dependent on the gametophyte and is the short-lived generation (the stalked capsule at the top of the leafy moss plant and the stalks on the liverworts and hornworts are the sporophytes).

They reproduce vegetatively (asexually) from small bits of haploid thallus tissue called gemmae, which are isolated in cups on the thalli and washed out by rain. Their sexual reproduction in which multicellular embryos are produced from fertilized eggs is restricted by the need for external water on the thalli surfaces in which the biflagellate sperms can swim to the egg.

Table 21-1 outlines some of the major similarities and differences between the three lineages of bryophytes.

Table 21-1: A Comparison of the Structure of the Three Bryophyte Lineages

Feature	*Liverworts*	*Hornworts*	*Mosses*
gametophyte	flat or "leafy"	flat	erect "stem" and "leaves"
stomata	no (have pores)	yes	yes
distinguish D-methionine	no	yes	yes
hydroids & leptoids	no	no	yes
axial gametophytes	no	no	yes
protonema	no in general	no	yes
"leaves"	1 layer, overlapping	none	1 layer, spiral, with midrib
chloroplast	many	1 large	many
symbiotic bacteria	some	yes	no
cuticle	no	yes	yes
determinate sporophyte growth	yes	no	yes

A Typical Bryophyte Life Cycle

Although individuals of the three bryophyte groups differ from one another morphologically and in other details, the moss life cycle shown in Figure 21-1 is typical of the group in general.

As do all plants, bryophytes alternate a gametophytic generation with a sporophytic one (a **sporic meiosis,** a life cycle in which meiosis gives rise to **spores,** not gametes). Each of the haploid ($1n$) spores is capable of developing into a multicellular, haploid individual, the **gametophyte.** The first structure formed from spores in most mosses and many liverworts is a filamentous, algal-like, green **protonema** (plural, **protonemata**). In some mosses the protonemata are long lived with rhizoids and aerial filaments and they often form dense green mats in suitable sites. Cells in the protonema, probably stimulated by red light and kinetin, give rise to shoots, which enlarge and become the mature gametophytes. In the bryophytes, these are the dominant, independent (photosynthetic) plants.

The gametophytes initiate **gametangia** on special branches or at the tip of the main shoot. In these structures the **gametes—eggs** and **sperms**—are produced during the sexual portion of the cycle. The female gametangium—called an **archegonium**—and the male **antheridium** may be produced on the same plant or on different plants. In both kinds of gametangia, a protective layer of non-reproductive tissue—a sterile layer—surrounds the inner reproductive cells. (A sterile layer is absent in algal gametangia and is considered an upward evolutionary step towards the protective seed coats of flowering plants.) Mature sperm, released from the tip of the antheridia when dew or rainwater is present on the surface of the plants, swim to the archegonia and down the necks to reach the eggs. One fuses with the single egg in each archegonium—the process of **fertilization**—thus combining the sperm and egg nuclear and cytoplasmic material. The resulting cell, a **zygote,** has a diploid ($2n$) chromosome number and is the beginning of the sporophytic generation. This reproduction is termed **oogamy**—a large, nonmotile egg is

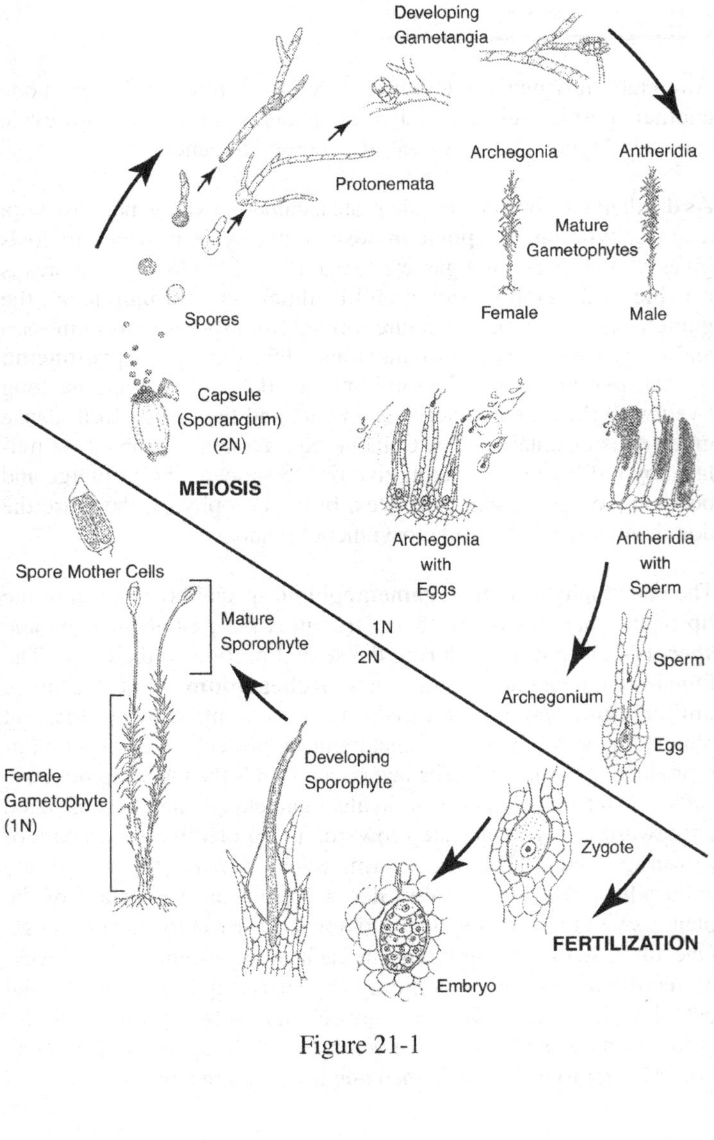
Developing
Gametangia
Archegonia
Antheridia
Protonemata
Mature
Gametophytes
Female
Male
Spores
Capsule
(Sporangium)
(2N)
MEIOSIS
Archegonia
with
Eggs
Antheridia
with
Sperm
Spore Mother Cells
Mature
Sporophyte
1N
2N
Sperm
Archegonium
Egg
Female
Gametophyte
(1N)
Developing
Sporophyte
Zygote
FERTILIZATION
Embryo

Figure 21-1

fertilized in the archegonium by a small, motile sperm that swims to the egg. In the bryophytes, an external film of water on the surface of the plant is the passageway for the biflagellate sperm; in more advanced plants, sperm move internally within special structures (pollen tubes) to reach the eggs.

After fertilization, the zygote remains in the archegonium and divides by mitosis repeatedly to form a multicellular, diploid **embryo,** the young sporophyte. Sugars and other materials are translocated from gametophyte to the developing sporophyte through **placental** tissue, a type of nutrition called **matrotrophy.** (No plasmodesmata connect the gametophyte and sporophyte; movement of material is along the cell wall, that is, it is *apoplastic movement*). The sterile jacket cells also divide and in mosses form a tight cap, the **calyptra,** over the tip of the developing sporophyte. The mature sporophyte in both liverworts and mosses consists of a **foot, seta,** and **capsule.** The moss capsule has modifications to assist in spore release: a cap, the **operculum,** covers the opening, and **peristome** teeth form a ring around the mouth of the capsule. Sterile cells, **elaters,** within the capsule are hygroscopic and as they alternately absorb water and dry out, they twist and turn pushing the spores upward and outward.

The hornwort sporophyte that develops from the zygote is an erect, long, green cylinder with an absorbing foot embedded in the gametophyte thallus. The sporophyte is photosynthetic and has stomata so it doesn't depend entirely upon the gametophyte for sustenance. Spores are produced in the cylinder around a central **columella** of sterile tissue and are released as the mature tip of the sporophyte dries out and twists in the air. At the base of the foot, a zone of meristematic tissue continues to divide and the sporophyte is thus continuously renewed from the base.

Phylogeny

With a sketchy fossil record, bryophyte phylogeny is based primarily on molecular sequencing of rRNA and morphology of extant species—

with hypotheses changing as new data accumulate. Researchers generally believe that the charophyte-coleochaete group of the green algae gave rise to the plants and that the bryophytes diverged from the common ancestor to vascular plants sometime during the Silurian, over 430 million years ago. Opinions differ concerning phylogenies within the bryophyte group, however. One view believes the three bryophyte groups had separate evolutionary origins and that only the mosses are more closely related to the plants. The bryophytes seem to have evolved during a time in which gametophyte characteristics were important for plant survival, whereas the vascular plants evolved when conditions favored the sporophyte portion of the life cycle.

Ecology

As descendants of the early land plants, bryophytes have retained the ability to adapt to a variety of habitats and environments. They live as understory components on the ground or as epiphytes in forests worldwide, but flourish most luxuriantly in moist warm-temperate and tropical habitats. Many of the liverworts and some species of moss are pioneers on newly burned ground while still other mosses colonize bare rock surfaces where their presence accelerates the erosion of rock to soil. The few desert dwellers rely on the condensation of dew on their surfaces to supply their metabolic water needs. In these sites, production of sporophytes is rare and the species spread primarily by vegetative means.

Mosses and liverworts are prominent in the arctic tundra on bare, dry surfaces while farther south in the circumpolar boreal conifer forest (taiga) and its southern extensions, the northern conifer forests, mosses constitute most of the biomass of the bogs and wet understory of the trees.

CHAPTER 22
THE PLANT KINGDOM: SEEDLESS VASCULAR PLANTS

The Ferns and Their Allies Lack Seeds but Have Vascular Tissue

The seedless vascular plants are intermediate in their structural and reproductive adaptations between the more "primitive" bryophytes and the "advanced" seed plants. They often are called the amphibians of the plant world for although their sporophytes are well-adapted to life on dry land, their gametophytes require a moist habitat to grow vegetatively and to reproduce sexually. The group includes the ferns and the "fern allies," the latter a collection of plants whose relatives were the dominant plants in Paleozoic landscapes for 60 million or more years. Today, the members are a few remnant species reduced to an exceedingly minor role in the flora. The fossilized remains of the early vascular plants exhibit a variety of ways of coping with the terrestrial environment—only some of which were successful.

The relationships among and within the groups remain unclear for three major reasons: 1.) fossils form an incomplete record, but are the basis for many of the conclusions, 2.) data from molecular RNA sequencing of living species are incomplete, and 3.) opinions vary among botanists concerning how the available morphological and molecular data fit together in proposed phylogenies. The prevalent hypothesis concerning the placement of the groups on the large tree of life distinguishes two main lineages of vascular plants that diverge very early in the development of a land flora. One line includes the most primitive taxa and the lycophytes, the other the ferns, horsetails, and seed plants. The "which-goes-where" disagreements primarily concern extinct groups, but some extant ferns are problematical in their relationships also.

On one level the possession of vascular tissue—xylem and phloem—separates the ferns and their allies from the bryophytes and the lack of seeds from the gymnosperms and angiosperms. Other characteristics they share in common are more varied and include:

- An alternation of a haploid gametophyte phase with a diploid sporophyte, a sporic meiosis. The gametophyte and sporophyte are nutritionally independent of one another.
- The sporophyte is the dominant, often branched, long-lived phase (the leafy fern plant is the sporophyte, for example). Many are perennial and vegetative (asexual) reproduction is common.
- The gametophyte is smaller and either photosynthetic or saprophytic. Because the flagellated sperm need water in which to swim to the egg (like bryophyte sperm), the gametophyte is restricted in distribution by habitat. The plants are oogamous.
- Eggs are produced in archegonia, one per archegonium; sperm in antheridia, many per antheridium. The gametangia are multicellular with a protective coat of sterile cells and borne in nearby areas on one gametophyte or on separate ones.
- Haploid spores are produced by meiosis in sporangia and in some of the seedless vascular plants are of two different kinds: microspores and megaspores. The sporangia develop on specialized leaves called sporophylls. Some members of the group have strobili (singular strobilus), cones in which the sporophylls are clustered.
- A cell plate separates the new daughter nuclei during cell division.
- Cell walls are cutinized (unlike the bryophytes in which a cuticle is lacking).
- Xylem and phloem are well developed and transport water, minerals, and carbohydrates throughout the large sporophytes.
- Cellulose is the common wall material; a secondary wall of lignin strengthens the cells of most of the group.

The living members of seedless vascular plants belong to four different phyla whose general characteristics are summarized in Table 22-1. Each group is described separately in the following pages.

Table 22-1: Characteristics of Four Phyla of Seedless Vascular Plants

Characteristic	*Lycophyta (lycophytes)*	*Sphenophyta (horsetails)*	*Psilotophyta (whisk ferns)*	*Pterophyta (ferns)*
common members	club mosses, *Lycopodium Selaginella*, quillworts, *Isoetes*	*Equisetum*, scouring rushes	whisk ferns, *Psilotum*, *Tmesipteris*	tree ferns, climbing ferns, *Polypodium*, maidenhair fern, bracken, *Azolla,Marselia*
spore type	*Selaginella, Isoetes* heterosporous; *Lycopodium* homosporous	homosporous; some fossil members heterosporous	homosporous	water ferns heterosporous; all others homosporous
leaves	microphylls	microphylls	none	megaphylls
steles	mostly protosteles	modified siphonostele	protostele	few protosteles; mostly siphonosteles (and complex modifications)
sporangia	on sporophylls	on sporangiophores in stroboli	lobed lateral; synangia	on sporophylls; many clustered in sori
body and other features	*Lycopodium* and *Selaginella* small, creeping plants; *Isoetes* a small hydroilica; phyte with sharp pointed linear leaves	jointed, ribbed, hollow stems impregnated with small leaves in whorls at the nodes; sporangia on special stems or terminal on vegetative ones	dichotomously branched aerial stem only, no leaves, nor roots; rhizomes with rhizoids anchor the plant; endomycorrhizae present in rhizomes	two kinds: eusporangiate and leptosporangiate; leaves develop with circinate vernation; small group of water ferns all derived from a terrestrial ancestor

Phylum Lycophyta—Club Mosses, Selaginellas, and Quillworts

The lycophytes are the oldest of the seedless vascular plants that have living representatives. They constitute one of the two major lines (clades) of vascular plants, which split probably in the Silurian Age, but at least by the Devonian. For the last 400 million years, therefore, they have developed independently from the rest of the vascular plants. During this time, they evolved from small, semiaquatic herbaceous plants to huge trees that dominated the Coal Age forests for 40 million years and then, as continental masses shifted and the climate dried, they declined in importance until most became extinct by late Carboniferous-early Permian time. Their structural features show convergence with taxa on the line leading to the flowering plants. Leaves, wood, trees, and reproductive structures that resemble seeds evolved in both lineages.

There are about 1,200 species today in three lycophyte families: Lycopodiaceae, Selaginellaceae, and Isoetaceae. Both of the latter two families have only one genus each—*Selaginella* with about 700 species and *Isoetes* with about 100. None of the lycophytes is over a meter or so tall, even in the tropics where they flourish and are the most abundant. Many are epiphytes growing high in the tree crowns. The temperate zone plants are small, trailing, evergreen plants that once were collected in quantity to place as crudely woven evergreen "blankets" on graves in cemeteries. Some *Selaginella* species are known as "resurrection plants" because they grow in arid sites and shut down metabolically during dry periods, rolling their aerial stems into tight balls and appearing lifeless. When moisture is available, they uncurl and flash green leaves into the sun, making and storing sufficient photosynthates to weather the next dry period. A number of lycopods are present in the arctic flora, and many form a groundcover on the forest floor in the northern and montane conifer forests.

The distinguishing features of the lycophytes, as shown in Table 22-1, are the arrangement of their vascular tissues and their leaves—microphylls with only a single vascular strand. The sporangia on the modern plants are kidney-shaped, like those of the ancestral forms, and borne on sporophylls clustered in strobili. A distinguishing ligule (scale-like outgrowth) is present in the *Selaginella-Isoetes* group.

The life cycles of members of the three groups vary. The lycopods are homosporous and the spores give rise to bisexual gametophytes, which in some species develop underground and live with the assistance of a mycorrhizal fungus; others develop on the ground surface. The gametophytic phase of the cycle may last several years (15 in some) before gametes are released and zygotes produced. The embryo develops slowly into the sporophyte, and the latter may remain attached and drawing sustenance from the gametophyte for a long period.

Selaginella species are heterosporous with two kinds of gametophytes. The megagametophyte (female gametophyte) develops within the megaspore, and when the spore wall breaks the archegonia are exposed. The microgametophyte (male gametophyte) develops biflagellate sperm, which also are released by breakage of the spore wall. They swim to the nearby archegonia; after fertilization the embryo develops within the archegonium, which is retained in the megagametophyte, a situation not unlike that in the flowering plants—except that no integuments and subsequent seed coat grow around the embryo of *Selaginella*. This is a forerunner of a seed, but not yet one.

The third group of lycopods, the *Isoetes* line, also is heterosporous with sporangia borne on the quill-like sporophylls that cluster around the corm. A distinctive feature of these plants is a cambium that produces secondary tissues and is located in the corm.

Phylum Sphenophyta—Horsetails

Only one herbaceous genus—*Equisetum*—of 15 species remains of this once large group of woody trees of Carboniferous Age forests. *Equisetum* is one of the easiest plants to recognize: It has jointed, ribbed and hollow stems impregnated with so much silica that a rasping noise is heard when stems are rubbed together. Another of its common names, "scouring rush", indicates one of the early settlers' uses of the plants. At each stem node there is a ring of small leaves fused in a sheath. Some species additionally have a whorl of branches at each node, which gives rise to the "horsetail" common name. The aerial shoots arise from an extensive rhizome system. *Equisetum* sperm—like those of the rest of the ferns and fern allies—require an external film of water in which to reach the eggs; *Equisetum* is most often found in sites that are moist for at least part of the growing season. Sexual reproduction is not necessary to propagate horsetails, however, which often spread vegetatively by means of rhizomes.

Equisetum grows worldwide except for Australia, New Zealand, and Antarctica. Most species grow in the Northern Hemisphere between about 40° and 60° N latitude. The modern plants resemble their arborescent ancestors that grew 250 million years ago—which might make *Equisetum* the oldest living vascular plant genus and the one least changed over time. The horsetails have no commercial role and are of minor importance in natural ecological systems.

The life cycle of *Equisetum* is basically that of the ferns (and the psilotophytes and some lycophytes) with only morphological differences. Sporangia are clustered in cones (strobili) at the tips of ordinary-appearing vegetative shoots in some *Equisetum* species, while in others the sporophyte is a simple non-green, short-lived, unbranched shoot with terminal strobili. The sporangia hang in groups of five to ten from umbrella-like sporangiophores that compose the strobili. Each spore is wrapped in four thickened, hygroscopic bands called elaters (but different in structure from the moss elaters). As the elaters dry, they twist and turn giving buoyancy to the spores. The spores germinate, producing small, green, thalloid gametophytes, which anchor with rhizoids to a moist surface rich in nutrients. Multiflagellate sperm swim in water to the archegonium, one fertilizes the single egg at its base, the zygote forms an embryo in the archegonium and the young sporophyte is nourished by the gametophyte to which it is attached until the organs are sufficiently developed to sustain the sporophyte independently.

Phylum Psilotophyta—Whisk Ferns

Two living genera, *Psilotum* and *Tmesipteris*, with only two species of the former and less than 30 of the latter, constitute the entire phylum. Both genera are weeds in the tropics and subtropics. *Tmesipteris* is confined to the islands of the South Pacific, including Australia and New Zealand whereas *Psilotum* is more widespread, reaching parts of the southern United States.

The sporophyte of *Psilotum* looks like a survivor from the Devonian age; it has no leaves nor roots, a protoxylem, a dichotomously branching green stem with small scales that bears bright yellow synangia (formed from three fused sporangia) on short lateral branches. It is homosporous and the spores develop into bisexual gametophytes that resemble pieces of sporophyte rhizome.

While many of the morphological and anatomical features fit, data from current RNA sequencing and other chemical analyses cause many botanists to reject the long held belief that *Psilotum* is a living ancestor of the seed plants. Instead, *Psilotum* appears to many botanists to be a *descendent* of the ferns by loss and simplification of structures. There are no known fossils of the Psilotophyta. So much for appearances.

Phylum Pterophyta—Ferns

The largest group of living seedless vascular plants—and probably the most familiar—are the ferns with about 12,000 species, over two-thirds of which are tropical. Ferns are an ancient group. Spores and leaf impressions of plants that lived 400 million years ago in the Middle Devonian have been found, but almost all of these early types (grouped simply as "preferns") were extinct by the Permian. The group to which most modern ferns belong, the Filicales, first appeared in the Lower Carboniferous, 300 million years ago.

Characteristics

The ferns are an extremely diverse group, and there is no single characteristic that defines them. The following features are present in most:

- Leaves, called **fronds,** are **megaphylls.** Most are **compound** with a rachis and numerous pinnae (or compound once again with pinnules). Almost all have **circinate vernation**—they are coiled (circinate) tightly in "shepherd's crook" or crozier fashion over the growing tips. These unroll as they mature (growth from the base to the tip like this is termed **acropetal**). The croziers are called **fiddleheads** and are eaten by some people, although many species are toxic.
- Stems, for the most part, are **rhizomes** that grow at, or just under, the ground surface. They have only primary tissues. "Tree" ferns have erect, thick trunks, the bulk coming from roots clustered around the small true stem. The more primitive species have a protostele, most have siphonosteles, and some have complex dictyosteles.
- Roots are simple, uncomplicated and arise adventitiously along the rhizomes near the base of the fronds.
- Sporangia are located, for the most part, on the undersides of ordinary leaves in clusters called **sori** (singular, sorus). In early ferns, and some living ones, sori occur on specialized, exceedingly unleaf-like leaves. In many ferns a small leaf outgrowth called an **indusium** covers each sorus. Two types of sporangia exist:
- **Eusporangia:** These sporangia are thick-walled and open by splitting transversely. They produce thousands of spores.
- **Leptosporangia:** These thin-walled, delicate sporangia are only one or a few layers thick. They have an area, the **annulus,** where cell walls are thickened. When the annulus cells dry out at maturity, the sporangium splits and, like a catapult, throws out the spores. (Spores are few—128 at most, but commonly 64.)

- Ferns are divided into two groups based on the kind of sporangium they possess. The more primitive are the eusporangiate, and the more advanced the leptosporangiate.
- Ferns have the highest number of chromosomes known in vascular plants.
- Most modern ferns are **homosporous** (two orders of water ferns and some extinct ferns are heterosporous).

Life cycle

The large, leafy fern sporophyte alternates with a small (3–4 mm), flat green gametophyte—called a **prothallus**—in the typical life cycle (see Figure 22-1, *Polypodium*, a leptosporangiate modern fern). The sporophytes of ferns are independent, divided into leaves, stems (rhizomes), and roots, and have vascular tissues whereas the gametophytes are small, photosynthetic thalli that live anchored to the ground with rhizoids. Many are heart-shaped and only one-cell layer thick. The gametangia are sunken or protrude from the underside of the gametophyte. Fern sperm have several flagella (hundreds in some species). When the sperm are released through a pore at the tip of the antheridium, they swim in a film of external water to the opening at the top of the archegonium and down the neck to the egg where fertilization takes place. The zygote divides within a few hours after fertilization and is supplied at first with nutrients (and perhaps hormones) through an absorbing foot attached to the gametophyte. A tiny sporophyte with a rhizome, adventitious roots along its surface, and a juvenile green leaf soon pushes out from under the prothallus, establishes independence, and the prothallus whithers and dies.

Ecology

Ferns, among the vascular plants, are second in number to the flowering plants and have adapted to all manner of habitats. Some are aquatic, some live in deserts or on dry rock cliffs, a few persist in the cold arctic desert, but mostly the modern ferns live in the tropics. Here the ferns attain some of the glory of their past. Some tree fern trunks exceed 70 feet in height, and their 15-feet-long leaves share canopy space with the angiosperm tree crowns. Tropical ferns grow vigorously as epiphytes on and over everything in the understory. The smallest ferns are aquatics that float on subtropical and tropical ponds with modified leaves a centimeter or less in size.

The First Vascular Plants

Scientists widely believe that the first land plants evolved during the late Ordovician to early Silurian, although fossils from this time are incomplete and difficult to interpret. By the end of the Silurian a land flora had evolved that throughout the next 50 million years of the Devonian (410 to 360 mya [million years ago]) continued to change, adapt to life on land exposed to air, and spread across a landscape previously devoid of vegetation. By the end of the period, small plants had given way to tree size, well-diversified vascular plants.

The lycophytes separated from the rest of the early land plants, evolved adequate reproductive, supportive, and transport systems, and, by the Carboniferous, were large swamp forest trees. Three groups of now extinct vascular plants were prevalent in Devonian times: the rhyniophytes, zosterophylls, and trimerophytes. The oldest known vascular plant is *Cooksonia*, a 6.5-centimeter-tall plant with dichotomously branched (forking into two) leafless stems with sporangia at their tips. Only bits and pieces have so far been recovered and no rhizomes or below ground parts have been found. It is a rhyniophyte and it and its relatives were extinct by mid-Devonian time.

The trimerophytes are the basal group of the lineage that gave rise to the flowering plants and are also the ancestors of the horsetails, ferns, and progymnosperms. Superficially, the trimerophytes resembled their rhyniophyte ancestors and the zosterophylls, but differed from them in bearing terminal sporangia on branch tips. At a meter or slightly less in height, these were the largest of the three groups of early land plants.

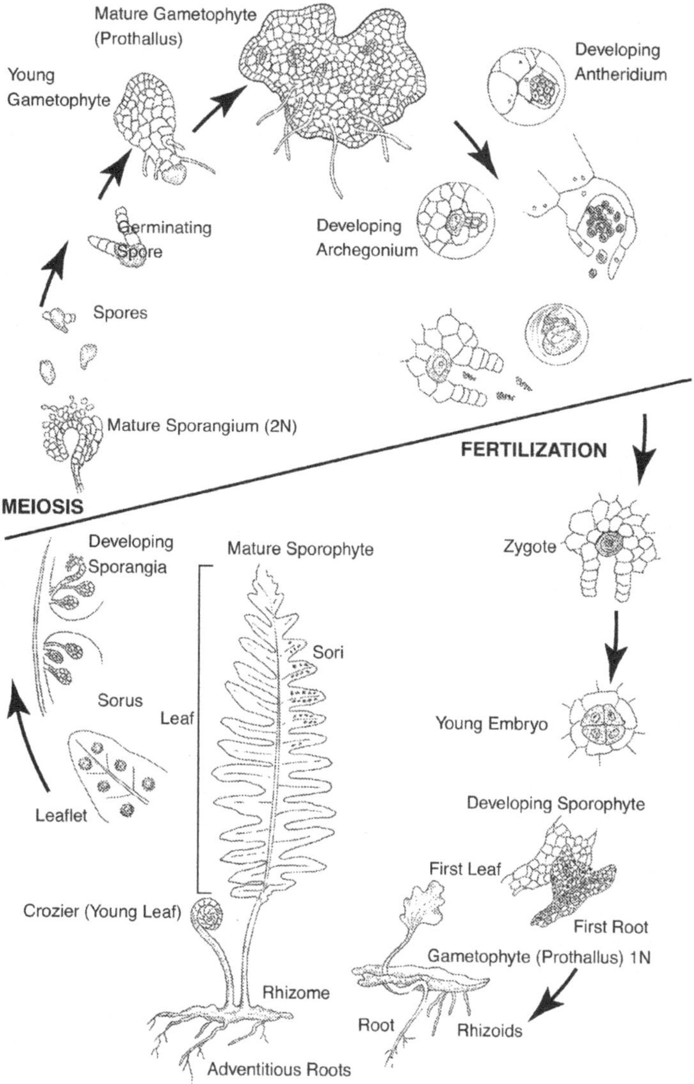

Mature Gametophyte
(Prothallus)
Young
Gametophyte
Developing
Antheridium
Germinating
Spore
Developing
Archegonium
Spores
Mature Sporangium (2N)
FERTILIZATION
MEIOSIS
Developing
Sporangia
Mature Sporophyte
Zygote
Sori
Sorus
Leaf
Young Embryo
Leaflet
Developing Sporophyte
First Leaf
Crozier (Young Leaf)
First Root
Gametophyte (Prothallus) 1N
Rhizome
Root
Rhizoids
Adventitious Roots

Figure 22-1

CHAPTER 23
THE PLANT KINGDOM: SEED PLANTS

The Most Successful Plants of All Time

Seed plants of today's world consist of two major groups, the "gymnosperms" and the angiosperms. Both evolved from a common ancestral group, the progymnosperms, during the Late Devonian period 365 million years ago. There are five phyla of seed plants with living representatives—four gymnosperms plus a single phylum of angiosperms, the Anthophyta. The fossils in several extinct phyla show evolutionary steps in the development of seeds and ultimately, the flowering plants.

The seed of the seed plants evolved from adaptations and changes to the megagametophyte, a reproductive structure common to all heterosporous plants (including the seed plants). The changes resulted first in the elimination of external water as the medium in which sperms reach the eggs, and secondly, produced a small, easily transportable package—the seed—to distribute the new sporophyte. Seeds give plants a tremendous advantage to spread and colonize new land.

The Gymnosperm Phyla

The four phyla of living gymnosperms are of separate clades or lineages, unlike the angiosperms, which are a monophyletic, single lineage. "Gymnosperm" means "naked seed" and the name draws attention to the ovules and resulting seeds that are exposed openly on the megasporophylls.

Phylum Coniferophyta

The conifers are woody, mostly evergreen trees, with needle-shaped or flattened leaves, which occupy the drier and cooler sites in the world today just as their ancestors probably did in the Permian. They are the familiar pines, firs, spruces, yews, hemlocks, and junipers of the Northern Hemisphere forests and the *Araucaria* species of the Southern. The *Sequoias* are among the tallest living trees (the Australians credit the *Eucalyptus*, an angiosperm, as being *the* tallest), and the bristlecone pines are among the oldest living plants. (*The* oldest plant, purportedly, is a clone of creosote bush in the Mojave Desert that is well over 4,000 years of age.)

The pines (pinus species). The leaves of the 90+ species of pines are needles that, in the seedlings, are borne singly along the stem. As the seedling matures, however, the needles appear in fascicles (bundles) of several (the number varying by species) on short shoots covered with scale-like leaves. The fascicle is a branch of determinate growth, a feature of evolutionary significance.

The pine needles are adapted for a xeric environment, one in which water is unavailable either because it is frozen most of the time or else because it is climatically scarce. Needles have a thick cuticle, an epidermis, and an underlying **hypodermis** of thick-walled cells, which further protect the mesophyll from drying out. The vascular bundles are surrounded by transfer tissue of parenchyma cells. **Resin canals** are present in regular patterns within the needle. Needles are shed at intervals of two to four or more years, but not all of the needles on the tree are dropped at the same time. Thus, a tree remains evergreen.

The wood of pines and conifers (called *softwood* by lumbermen) in general lacks vessels and is composed of tracheids with circular bordered pits. Parenchyma is almost entirely restricted to ribbons of narrow rays. The vascular cambium is bifacial; that is, it produces secondary xylem (wood) toward the center and secondary phloem toward the outside of the stem.

The life cycle of a pine is a slow, two-year process (Figure 23-1). Pollination occurs in the spring of one year, and the pollen tube begins its growth towards the megagametophyte about this time although its goal, the egg, is not as yet differentiated. In fact, the megaspore mother cell has not yet divided, the megagametophyte does not exist, and there are no archegonia, let alone eggs within the ovulate cones. This seeming lack of syncronization is of little concern because it takes the pollen tube over a year to digest its way through nucellular tissues to the archegonia—which gives ample time for megagametophyte preparations—and for the immature male gametophyte (the four-celled germinated pollen grain) to produce two sperm cells by division of the generative cell.

In the spring of the year following pollination, events come together: The eggs in the two to three archegonia are fertilized (polyembryony), and development of the new sporophytic generation begins. Normally only one embryo survives to maturity in the seed. The pine seed consists of tissues from two sporophyte and one gametophyte generation. That is, the parent $2n$ sporophyte tissue remains in the seed as the seed coat (mature integuments); the embryo is the new $2n$ sporophyte, which is surrounded by the $1n$ megagametophyte. Sometime during the summer of the year following pollination the seed matures and is shed. If it germinates in a suitable habitat, a new tree grows to initiate still another cycle.

Other conifers. Not all of the conifers resemble the needle-leaved pines in appearance or length of time to complete the sexual reproductive cycle—most take only a year. Some conifers are deciduous, such as larch (*Larix*), bald cypress (*Taxodium*), and the dawn redwood (*Metasequoia*). The yews (*Taxus*) have flattened leaves and instead of a cone have a fleshy red cup, an aril.

As a group, the conifers occur throughout both the Northern and Southern Hemispheres, some in large numbers and widespread across sites largely unsuitable for angiosperm tree growth. Many other genera are restricted in species numbers and have a relict distribution. Some, such as the dawn redwood and the Wollemia pine, have only a few living individuals left in isolated sites although botanists knew of neither species until fairly recently.

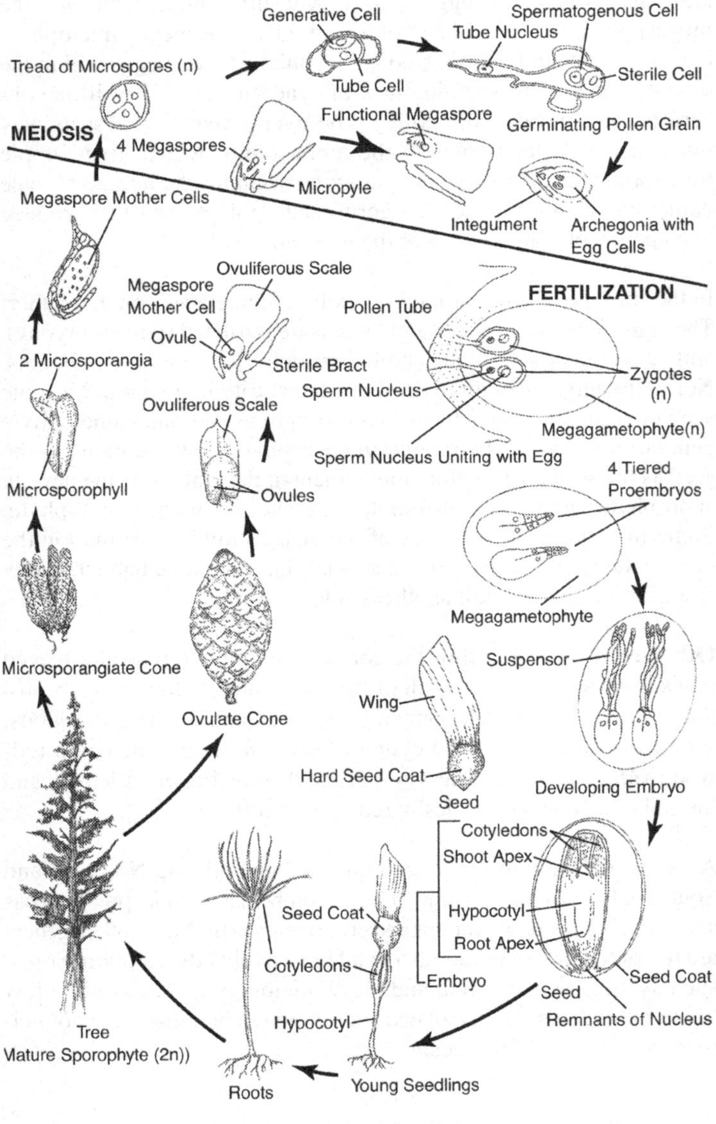
Generative Cell
Spermatogenous Cell
Tube Nucleus
Tread of Microspores (n)
Sterile Cell
Tube Cell
Functional Megaspore
Germinating Pollen Grain
MEIOSIS
4 Megaspores
Micropyle
Megaspore Mother Cells
Integument
Archegonia with Egg Cells
Ovuliferous Scale
Megaspore Mother Cell
FERTILIZATION
Pollen Tube
Ovule
2 Microsporangia
Sterile Bract
Zygotes (n)
Sperm Nucleus
Ovuliferous Scale
Megagametophyte(n)
Sperm Nucleus Uniting with Egg
4 Tiered Proembryos
Microsporophyll
Ovules
Megagametophyte
Suspensor
Microsporangiate Cone
Wing
Ovulate Cone
Hard Seed Coat
Developing Embryo
Seed
Cotyledons
Shoot Apex
Seed Coat
Hypocotyl
Root Apex
Cotyledons
Seed Coat
Embryo
Seed
Hypocotyl
Remnants of Nucleus
Tree
Mature Sporophyte (2n))
Roots
Young Seedlings

Figure 23-1

Seventeen species of conifers are growing in small numbers along the California coast—and nowhere else. The same story is repeated in other genera, which leads to the conclusion that the conifers reached their heyday in the Mesozoic and then started a decline that continues to the present.

Other Gymnosperm Phyla with Living Representatives

The surviving gymnosperm are a diverse group that persist today in restricted habitats or in regions too extreme—too hot, too dry, too cold—for angiosperms. Some preserve in their structures and life-styles evolutionary early "flowers" that didn't quite succeed and the experimental lifestyles later adapted and adopted by the angiosperms.

Phylum Cycadophyta

The cycads look like palms with cones and are prevalent worldwide in the tropics and subtropics (two species grow wild in Florida). Some grow 15 meters or more in height, but many have shorter trunks and an almost rosette appearance. The ovulate cones are large (some weigh over 30 kilograms) and are borne upright on megasporophylls among the vegetative leaves. Pollen and ovulate cones are produced on different plants. Despite the un-pine-like appearance, the life cycle of the cycads is similar to the pines. The sperm, however, though carried to the archegonia in a pollen tube, are multiflagellate with hundreds of flagella.

Zoologists consider the Jurassic Period the "Age of Dinosaurs," but botanists refer to it as the "Age of Cycads." Cycads and bennetti-taleans, an extinct group of plants that often are mistaken for cycads because of the close resemblance of their leaves and growth form, dominated the land flora.

Phylum Ginkgophyta

Another un-pine-like gymnosperm is *Ginkgo biloba*, the maidenhair tree, the sole remaining representative of a group of important plants of the Mesozoic forests. It has broad, fan-shaped leaves with dichotomously branching veins and is deciduous. It is widely planted as a street tree because it withstands well the air pollution of cities. Its seed coats, however, have a foul odor, and when the seeds fall around the trees and rot in the heat of summer, it becomes a much less desirable plant. There are no known wild *Ginkgos*; the plants of today derive from stock preserved in temple gardens by monks in China and Japan. They distributed seeds to gardeners around the world over 200 years ago.

Several features unite *Ginkgo* with the rest of the gymnosperms, but to which precise lineage—the conifer line or the cycad—is still being debated. No cones are produced and the female gametophyte is contained in a cherry-like seed. The ovules and microsporangia are produced on different trees.

Phylum Gnetophyta

Three living genera—none of whom resemble one another or any other living gymnosperm—constitute the Gnetophyta, *Gnetum, Ephedra,* and *Welwitschia*. The gnetophytes are the closest living relatives of the flowering plants, and they form a monophyletic clade. *Gnetum* species are tropical vines and trees that resemble flowering plant species with their broad, simple leaves. *Ephedra*, called joint-fir or Mormon tea, is a desert shrub with worldwide distribution. The species of *Ephedra* have green, jointed stems and small scale-like leaves. They produce several secondary metabolites that are chemically similar to human neurotransmitters, and people have used the plants as medicinal teas for centuries. *Welwitschia*, among the weirdest of plants, is confined to the Namib Desert of southwestern Africa and has a buried trunk on which two strap leaves of indeterminate growth are attached. The leaves split lengthwise into strips and blow about on the shifting sand. Ovulate and pollen cones are produced on separate plants on the rims of the exposed trunks. The eggs in the megagametophyte move towards the pollen tubes in their own tube-like structures; fertilization takes place after the two tubes fuse.

There are few good gnetophyte fossils, the best evidence of their past occurrence being pollen that resembles that of *Ephedra* found in Triassic and Cretaceous strata. This peculiar collection of plants has several angiosperm characteristics, but none of the plants in the group is the direct ancestor of the angiosperms.

Extinct Gymnosperm Phyla

The fossil record of ancient gymnosperms is surprisingly complete and provides good data from which to construct phylogenies.

Phylum Progymnospermophyta

Many botanists believe the progymnosperms are the most likely ancestral group from which the seed plants evolved. They have many features of the seed plants, but are still spore producers and are not themselves seed plants. They were important members of the vegetation from Middle Devonian through the Lower Carboniferous. Many were large trees with fern-like leaves and probably formed forests in these early landscapes.

One feature of evolutionary significance that advances and separates the group from the ferns and trimerophytes is the bifacial cambium present in the progymnosperms; the group had a vascular cambium that produced secondary xylem and secondary phloem. A bifacial cambiumis is a characteristic of the seed plants and it appears for the first time in this group. The fossil progymnosperm wood resembles that of more modern conifers with tracheids and bordered pits.

Phylum Pteridospermophyta

Another Late Devonian group is the seed ferns, with leaves so like ferns that if no seeds are attached the fossils often are cataloged as ferns. This is an unnatural mixed group (like the protista) with no taxonomic ranking and not enough specimens to determine phylogenies well. One hypothesis derives the seed ferns from the progymnosperms in a lineage with no modern descendants. Another places one of the seed fern lines, the Medullosans, in a clade with the cycads—with good arguments for and against the arrangement. The group remains a puzzle.

Phylum Cordaitales

This phylum derives its name from an extinct genus of trees that formed upland and swamp forests in late Carboniferous and Permian time. Current understanding of the phylogenies of the group place *Cordaites* near the cycads, but their lineage is still a matter of much discussion. They had long strap-like leaves and produced two kinds of cones, pollen and seed, on separate branches. They had a well-developed root system with secondary xylem. The stems also had secondary xylem surrounding a large central pith.

Phylum Bennettitales

As mentioned earlier in this chapter, the bennettitaleans resemble modern cycads and for many years were placed in the seed fern melange. Some recent phylogenetic analyses—aimed at finding the origin of the angiosperms—place them in or near the angiosperm clade with ancestory still unknown. One bennettitalean, *Williamsoniella*, from the Jurassic, had a bisexual strobilus—a rare feature in gymnosperms that mostly have separate organs for pollen and ovules. The microsporophylls were in whorls surrounding a center of several ovulate whorls and the whole was enclosed in bracts. In many ways, this is the structure of a flower before there were angiosperms on the scene.

Phylum Anthophyta—The Flowering Plants

The Anthophyta, the angiosperms or flowering plants, is the largest and youngest phylum of plants and the one whose members dominate the vegetation of the modern world. The origin of the angiosperms is an enigma, but from all evidence they probably arose sometime during the Late Jurassic or Early Cretaceous, but were first fossilized in the Cretaceous. As they evolved over the ensuing millenia, they developed adaptations that made them ever more successful in the competition for a place in the sun. At the same time the angiosperms were diversifying, the terrestrial animals, too, were diversifying and taking advantage of the new food source. The extraordinary success of the angiosperms is not just that they have flowers—which certainly helps—but because they have, as well, a combination of other structural, developmental, and ecological features not found in their entirety in other groups.

The short list of characteristics of angiosperms includes:

- Flowers, the means of reproduction.
- Ovules surrounded by two integuments.
- Double fertilization, which leads to formation of polyploid endosperm tissue.
- Simple microgametophytes and megagametophytes. The microgametophyte is a three-nucleate structure, the megagametophyte an eight-nucleate one.
- Stamens with two pairs of pollen sacs.
- Sieve tubes and companion cells in the phloem, vessels in the xylem.

Evolution of flowers. With a spotty, incomplete fossil record of the early flowers, much of the understanding of flower evolution is inferred from modern flowers. Taxonomists for a century have defined angiosperm families on floral structure and separated "primitive" from "advanced" features. In this assessment, early, primitive flower characteristics are: an undifferentiated perianth with sepals and petals alike and separate; an indefinite number of parts in each floral whorl; spiral attachment superior ovaries; radial symmetry; and so forth.

Early carpels were leaf-like and seeds were borne on the edges. In advanced flowers, the carpel is folded inward and the seeds are enclosed. Closed carpels have differentiated stigmas, styles, and ovaries. The pollen does not land on the ovules directly.

Pollination. Flowers and their pollinators coevolved; that is, two or more species act as selective forces on one another and each undergoes evolutionary change. Early flowers probably were wind pollinated, but the selective advantages of cross-fertilization by animal pollinators must have been a powerful selective evolutionary force from the very beginning

Specializations to ensure cross-fertilization and attract pollinators include: colors in wavelengths visible to the pollinators; nectaries placed so that access requires passage across pollen sacs; odors; structural changes such as long corolla tubes and spurs filled with nectar.

Dispersal. Concomitant with the changes to insure fertilization are those that insure dispersal of the products of fertilization, such as the seeds and fruits. Fruits can be dry or fleshy, remain closed or split open at maturity, have hooks or spines that attach to fur or feathers. Seeds can have hard coats, colors, wings, plumes, and all manner of other clever ways to move the new generation away from the old—which is the underlying point of the whole process. Dispersal not only permits colonization of new areas by a species, but also prevents competition for water and minerals between parent and offspring at the home site.

Secondary metabolites (products). Chemical compounds produced by plants are either: 1.) primary products found in all plant cells that are necessary for life, such as amino acids, or 2.) secondary products found in some cells that are important for the survival or propagation of the plants that produce them. When the secondary products were first discovered they were thought to be waste products that plants neither were able to utilize nor get rid of so they were stored out of the way in the vacuoles. With further research it became apparent that the materials were not simply wastes, but had a purpose—to ward off insect attacks, to stop herbivores from eating the plants, or as a response to bacterial and other pathogens.

The toxicity of many of the products is not confined to insect attackers; humans who consume the plants also are affected. Alkaloids produced as secondary metabolites include: cocaine, caffeine, morphine, nicotine, and atropine—a potent pharmacological arsenal. Terpenoids are another class among which are the hydrocarbons, which plants release from their leaves in prodigious amounts and which contribute to air pollution. Terpenoids form the haze that makes the Great Smoky Mountains "smoky." Rubber is a terpenoid as are taxol and menthol; so are the carotenoids of the plastids and sterols of the cell membranes. Phenolics are imporant secondary metabolites whose plant roles are still being discovered. The evolution of secondary metabolites gave flowering plants a biochemical means to cope with the environment—and added still another improvement over their neighbors.

Systematics

Phylogeny. There are more questions than answers in the phylogeny of the angiosperms. Part of the problem lies in the lack of an adequate fossil record. The first clearly angiosperm fossil is from the Early Cretaceous and is an impression of a fully developed flower. Molecular RNA/DNA sequencing currently is being applied in new phylogenetic (cladistic) analyses to answer the question of angiosperm origins. As yet, no generally accepted answer exists, but several hypothesis are being hotly debated. The molecular data indicate the seed plants most closely related to the angiosperms are the gnetophytes and bennettitaleans, which, incidentally, is the same conclusion reached by some botanists using morphological and anatomical features 50 years ago. Others at the time favored the "seed ferns" as angiosperm ancestors. A second debate revolves around the nature of the first angiosperms. Were they woody or herbaceous? There are no clear answers in that debate either. Cladists in general favor a woody origin, but there are equally vociferous advocates for the herb hypothesis.

Classification. The long-held separation of the angiosperms into two groups on the basis of the number of cotyledons in their seeds—monocots (one) and dicots (two)—is an artificial classification now being abandoned in favor of one based on molecular data, which recognizes evolutionary relationships. The 235,000 species of angiosperms are separated into three groups:

- **Eudicots:** 165,000 species; two cotyledons, leaves with net venation, primary vascular bundles in a ring, vascular cambium with secondary growth, pollen with three pores; flower parts primarily in fours or fives or multiples of four or five.
- **Monocots:** 65,000 species; one cotyledon, leaves with parallel venation, primary bundles scattered, vascular cambium rare, pollen with one pore; flower parts in threes or multiples of three.
- **Magnoliids:** 5,000 species; primitive characters, pollen with one pore, cells with ether-containing oils; two subgroups: woody magnoliids and paleoherbs; most have fused carpels.

The monocots are a monophyletic group with a common ancestor based on their single cotyledon and a few other features. So, too, are the eudicots with their triaperturate pollen. The magnoliids, however, have no uniting feature and their evolutionary relationships are still being worked out.

CHAPTER 24
ECOSYSTEM STRUCTURE AND FUNCTION

The Parts of Ecosystems

Plants originate, live, and die in **communities** consisting of plants and a variety of other organisms, large and small. All the living constituents make up the **biota** (or the biotic component) of the community. The nonliving, physical factors (such as temperature, light, and nutrients) that affect the life of organisms—and determine to a large extent the kinds of organisms present—collectively represent the **abiotic** factors of the community. The **environment** is the total of all the biotic and abiotic forces in which an organism or a community of organisms lives.

The biotic components and abiotic portions of the environment are inseparably intertwined and constantly exchange materials; there are **inputs** (materials or organisms entering) and **outputs** (materials or organisms leaving) to the community; **energy** flows one way *through* the community, **nutrients** cycle *within* it. The community functions, therefore, as a dynamic system, an **ecosystem.** The ecosystem is the most complex level of biological organization.

The ecosystems of the world are distributed in environmentally controlled patterns or **biomes,** which contain distinctive communities with similar **life forms** (but not necessarily the same species) across broad geographic areas. *Deserts* are one kind of biome, for example. Chapter 26 discusses the terrestrial biomes. Freshwater and marine aquatic ecosystems are not categorized in this manner and are not discussed further in this book.

Ecologists are biological scientists who study the relationships of plants and animals to one another and to their environment. **Ecology** attempts to understand where, how and, if possible, *why* organisms live where they do. It searches for theories that can be formulated into ecological principles to better understand the place of organisms in the world ecosystem. **Environmental science** is an applied science that addresses the problems created by human activities on the functioning of the environment and searches for ways to alleviate the stress created by a burgeoning human population. It draws upon knowledge from all areas in a quest for solutions. Pollution of the environment, a depletion of resources, and an increasing need for food and energy are subjects of major concern to environmental scientists.

Ecosystem Structure—Plant Communities

The plant community usually is the largest visible part of an ecosystem, and often both the community and the ecosystem are named for the **dominant** plants present—that is, the plants that, by virtue of their size or numbers, modify and control the environment. The community is not a haphazard collection of organisms, but consists of **populations** of individuals whose **tolerance ranges**—the range of environmental conditions in which individuals of a particular species will grow—match those of the site.

Succession

Even the most stable ecosystems are in constant, normal flux. One of the easiest ecological processes to observe is **succession,** the change in the composition of the vegetation of a particular site over time. Two kinds occur. **Primary succession** takes place on newly exposed surfaces such as might appear after a volcanic eruption or following a rockslide in the mountains. **Secondary succession** occurs when vegetation is removed from land and new kinds of plants return to colonize the bare ground. On sites undergoing primary succession, no soil is present, and vegetation and soil develop concurrently. In secondary succession, vegetation develops on soil already in place, but the soil changes over time as the new colonizers develop new communities above, and below ground.

Weedy annuals are the usual **pioneer species** that colonize bare ground. They are "generalist" species with broad tolerance ranges and by their growth change conditions at the site, making possible the development of communities of other species. Productivity in the early stages is high, but as species richness and total biomass increase, productivity decreases (the reason why agricultural ecosystems are kept in an early successional stage).

Disturbance

Disturbances are necessary to sustain ecosystems and are the mechanisms by which diversity in kinds and ages of species and habitats is maintained. Some disturbances are large scale and rare (a tornado or hurricane, for example), but most are small and frequent (the blowdown of a few trees opening a hole in the forest canopy or rodent burrowing that brings underlying soil to the surface and destroys existing ground cover).

Ecosystem Functions

Three levels of organisms regulate the flow of energy in ecosystems: the **producers,** the **consumers,** and the **decomposers.** They are organized in complex **food webs. Autotrophs**—plants, algae, and some bacteria—are the **primary producers** of an ecosystem. **Heterotrophs**—animals, fungi, most protists and bacteria, and a few non-green plants—are the **consumers** in ecosystems. They obtain their energy and carbon from the organic material produced by the autotrophs. Four **trophic** (feeding) levels are recognized: The primary producers constitute the first level, followed by three levels of consumers. **Primary consumers** are the **herbivores** (plant eaters) that feed directly on the primary producers. The next level includes flesh-eaters, the **primary carnivores** that consume the herbivores. The top or fourth level is that of the **secondary carnivores** that dine on the primary carnivores. At each level, some of the energy acquired is used to do the metabolic work of the consumer, some is stored within the substances of the consumer's body, and much is lost to the environment as heat (not really "lost" since the heat maintains the temperature balance of the Earth and drives the wind circulation patterns that produce the climates).

Another group of organisms vital to the ecosystems are the **decomposers** that receive energy from all levels (and may contribute energy to some). Because they live on the detritus of ecosystems they sometimes are referred to as **detritivores.** Without their activities, minerals would not cycle through the biosphere, but would remain locked in the bodies of the organisms that got to them first, and life on Earth would cease—quite an important position for a group of organisms too small to be seen with the naked eye who live buried in the ground.

Element cycling

Two separate types of cycles keep elements moving through ecosystems: **gaseous cycles** in which the atmosphere is the reservoir and **sedimentary cycles** in which the rocks of the Earth's crust are the reservoir. Chapter 25 discusses cycles of five elements important to life.

Plant Interactions with Other Organisms

Ecology is the study of interactions of organisms with one another as well as with their environment. Plants, with their sedentary existence and need to attract pollinators or prevent herbivores from consuming them whole (because they can't run away from them), have evolved a different set of behavior patterns than have animals.

Competition

Competition results when an individual plant interferes with the needs of another plant for the same environmental resource (such as light, minerals, space) or when members of one population interfere with members of another for the same environmental resource. In plants, competition generally is *indirect,* through the resource, not *direct,* one-on-one (plants don't engage in leaf-to-leaf combat). Plants with the same life form and growth requirements are often in competition but surviving in slightly different microenvironments. This
generally leads to a better utilization of the resource and, with natural selection in operation over time, a greater diversification of the community.

Allelopathy

Allelopathy is a particular form of *direct* competition in which one plant species (or a fungus like *Penicillium*) produces a substance toxic to another. In some instances, the substance inhibits the development of the producer's own seeds or spores. The compounds may leach from the roots into the soil or accumulate in the ground around the plant as leaves drop and decay. Some are terpenes that volatilize and are spread through the air as aerosols. The essential oils of members of the mint family are toxic to numerous plants, as is the oil of black walnuts. Caffeine produced by tea and coffee plants inhibits the growth of seedlings of many species.

Secondary metabolites

Chemical warfare of another kind is waged by plants that produce **secondary metabolites**—chemical substances that protect the plants from being eaten by herbivores. Plants and their predators undoubtedly coevolved, with changes in one instigating reactions and further evolutionary changes in both.

Some of the metabolites are not merely deterrents, but are chemicals that imitate hormones, enzymes, or other essential compounds of animal bodies. One metabolite interferes with insect metabolism by inhibiting the juvenile growth hormone. Others, like the alkaloids morphine and cocaine, affect the human nervous system; and caffeine, although a stimulant to humans, in plants is toxic and lethal to insects and fungi. The estrogens produced by some plants have no known role in the plants, but their importance to human reproduction is well known—and a cause for concern when humans eat vegetables.

Defense substances of a different kind protect plants from bacteria and fungi attacks. These substances, called **phytoalexins,** act as natural antibiotics and protect the plant from bacteria and fungal pathogens when leaves are damaged or stems wounded. Nicotine in tobacco plants is synthesized in response to wounding.

Symbiosis

In a **symbiosis,** two different kinds of organisms live together in an intimate and more or less permanent relationship. *Lichens* are the classic example of a symbiosis between a fungus and a cyanobacterium or an alga. *Mycorrhizae*, too, are examples of fungi and the root cells of vascular plants in a symbiosis. If the interactions between the symbionts are of mutual benefit, the symbiosis is termed a **mutualism;** if one partner benefits and the relationship is of no significance to the other, it is a **commensalism; parasitism** is a symbiosis in which one partner benefits and the other is harmed.

Mutualism. Seed plants have developed all manner of mutualisms, the most highly developed being the interactions between insects, birds, bats, and a few other animals that ensure pollination of flowers, especially by cross-fertilization. Pollinators are attracted to the flowers by colors, scents, and nectars and once on-site, all manner of structural floral adaptations insure the pollinator gets a dusting of pollen to take to the next flower it visits. The pollinator gets food, and the plant gets a messenger service more effective than chance winds.

Seed and fruit dispersal mechanisms also are well-developed, coevolved mutualisms. Succulent edible fruits with their scents and colors are great dispersal devices geared to larger animals and often found on plants that produce seeds with hard seed coats. The coat may be so difficult for water to penetrate that germination is not possible unless some mechanical abrasion or chemical solvent is applied. The gizzard of birds is an effective grinder, and the stomach acids of mammals take off much of the seed coat before hard-coated seeds are expelled in the feces.

Parasitism. Bacteria, viruses, and fungi have not spared the plants as hosts for their parasitic lifestyle nor have vascular plants that parasitize other vascular plants. The lines among mutualism, commensalism, and parasitism are often blurred because the definitions are based on value judgments, that is, on the degrees of harm or benefit to the symbionts. About 3,000 species of vascular plant parasites are worldwide in their distribution. Some of these have lost the ability to photosynthesize entirely, but others attach to the vascular system of their hosts and divert the water and minerals in transit to their own photosynthesis.

CHAPTER 25
BIOGEOCHEMICAL CYCLES

Materials Used by Organisms are Recycled

The Earth is essentially a closed chemical system through which the elements necessary for life are reused and move from abiotic **reservoirs** to the biota and back in global **biogeochemical** cycles. Some elements are held as gases in the atmosphere, others are components of the lithosphere (rocks and soil of the Earth's crust), many move through the hydrosphere (marine and freshwaters) before or after their sojourn in the biosphere (the living components).

The cycles through the lithosphere are said to be **sedimentary** cycles (from the sedimentary rocks in which the elements reside) and are of such long duration that the elements are essentially removed from further cycling until tectonic (mountain building) or volcanic eruptions expose the rock layers to new weathering. Elements have shorter residence times in the air in the **atmospheric** cycles and generally the least of all in the biota. A surprisingly small amount of the world's matter is held in living organisms at any time; the reservoirs for the elements of concern to life are almost entirely abiotic ones.

Plants are more than merely users of the chemicals of the Earth; through their metabolic processes they exert a considerable influence on the cycling of the major chemicals. Plants have been indispensable through **deep (geologic) time** in maintaining the steady-state condition of most of the biogeochemical cycles. All of the ecologically significant chemical elements have both an abiotic and a biotic component. Carbon, hydrogen, and oxygen enter plants from the air and from the decomposition of organic matter, but the other 14 essential nutrients are taken from the soil, as are the miscellaneous other elements used in small amounts by a variety of organisms. Nutrients released from weathered rocks enter the soil solution and move by **diffusion** and **mass flow** to the sites of biological activity. **Rock weathering** is a long-term process that adds small quantities of minerals slowly, over time to the ecosystem. Plants and other organisms, therefore, obtain most of the minerals they need by recycling existing organic matter.

This chapter discusses the elements separately, but in reality no cycle operates alone; all are intertwined and dependent upon one another.

The Water Hydrologic Cycle

All life depends on water and in its absence life ceases. The kind of vegetation present at a site depends upon the amount of free water available and a principal factor in terrestrial net primary production is the amount of precipitation a site receives. The movement of water on the face of the Earth affects the rate of weathering, the amount of materials carried to the seas and lost to the bottom sediments, the erosion of soils, as well as the global heat balance and rainfall patterns from pole to pole.

Water is the carrier of the elements, and all of the biogeochemical cycles include portions of time in the hydrosphere. Water itself cycles and would do so in the absence of organisms (unlike the other major elements, which require organisms in their cycles). Oceans, which cover over three-quarters of the Earth's surface, are the reservoir for water. Evaporation from their surfaces amounts to 425,000 km^3 per year. About 90 percent of the water returns to the oceans as precipitation, and the remaining 10 percent falls as precipitation on the land. Transpiration (the loss of water from plants) and evaporation from the soil, together called **evapotranspiration,** add 71,000 km^3 to the atmosphere yearly. The amount of water in the atmospheric reservoir is small, however, and the terrestrial water reserves are in the **groundwater,** the lakes and streams, and the soil water. It is estimated that 96 percent of the freshwater in the United States is held in the groundwater reservoir. Groundwater doesn't enter the global biosphere directly and is unavailable to plants except through human activities that remove water in wells drilled into the supply. Recharge of the reservoir is from seepage through the soil and is a slow process.

Of particular significance is the unevenness of the water cycle across the Earth. Some regions receive quantities of rain, others none. Nor is evaporation from the oceans uniform. Near the equator evaporation may average 4 mm/day and in polar latitudes <1 mm/day. Because solar energy powers evaporation and is not uniformly received due to the tilt of the Earth toward the sun, the heat balance of the Earth depends on the transfers of heat through evaporation and precipitation of water.

The Carbon Cycle

Carbon—the element that defines life—enters the biota through photosynthesis while the oxygen released in the process makes possible aerobic respiration of all living things. Molecules that contain carbon are the major constituents of living tissues, but the amount of carbon in active biosphere cycling is minor compared to the amount held in abiotic reservoirs such as sedimentary rocks, fossil fuel deposits, and deep sea sediments.

Respiration and photosynthesis are the driving forces of the carbon cycle. Carbon enters the biosphere as atmospheric carbon dioxide (CO_2), which is incorporated by photosynthetic organisms into carbohydrates. It leaves, also as CO_2, through the respiration of organisms. The carbon cycle is thus bound with that of oxygen and, through oxidation-reduction reactions, to other elements of importance to organisms. Removal of CO_2 from the atmosphere by terrestrial vegetation in photosynthesis is balanced by the return due to respiration of plants and decomposer organisms in the soil. Plankton organisms in the surface waters of the oceans also remove and return atmospheric carbon in about the same proportions, although a minor portion of dissolved CO_2 is lost in the deep ocean and buried in the sediments. The atmosphere links all compartments in the cycle, and the changes in atmospheric CO_2 are a measure of the health of the ecosystem.

A disturbing feature of the current carbon cycle is the net atmospheric increase of CO_2. In the last 50 years of accurate measurements, annual increases of about 0.4 percent (1.5 ppm) have occurred consistently—indicating that a cycle in place for the last 100,000 years no longer is in balance. An increase of as much as 30 percent in the last 200 years (since the beginning of the Industrial Revolution) is circumstantial evidence that the burning of fossil fuels together with human alterations to the natural vegetation cover of large areas of the globe may be the cause. It is suspected that the increased levels of CO_2 will trap more reflected radiation in the atmosphere and increase the temperature on the surface of the Earth causing a *greenhouse effect.*

The Nitrogen Cycle

The atmosphere holds the greatest reservoir of nitrogen, but, as gaseous, triple-bonded N_2, it is chemically inert and unusable by plants and almost all other organisms. A few kinds of bacteria that possess the enzyme **nitrogenase** are the exceptions. They are able to convert (reduce) N_2 to ammonium ions (NH_4^+), which many organisms, including plants, are able to metabolize. The process is called **nitrogen fixation** and ranks equally with photosynthesis in significance to life. When organisms die, decay bacteria and fungi release the fixed N of organic compounds. New organisms then reformulate it into amino acids—hence proteins—together with nucleic acids, nucleotides, coenzymes, and vitamins all of which are essential to life. The **denitrifying** bacteria return some nitrogen to the atmosphere as N_2. This process—**denitrification**—is anaerobic and takes place in almost all types of soil. Microorganisms, therefore, control the major phases of the N cycle.

There is, in addition, an abiotic part to the cycle. A lesser amount of nitrogen is fixed from the atmosphere through reactions occurring in the gaseous emissions from volcanoes and from lightning discharges.

This nitrogen is washed out of the atmosphere by rain and is added to the soil and water where it, too, can be used by organisms.

Human activities have a significant effect on nitrogen cycling. Production and use of nitrogen fertilizer, combustion of fossil fuels, and planting crops that fix nitrogen have unbalanced the previously stable relationship between fixation and denitrification. Gaseous industrial pollutants—abbreviated as NO_x compounds—foul the air in many cities and wash out in sufficient amounts to constitute "acid rain" in some parts of the industrialized world.

The nitrogen cycle can be understood most easily by looking at its separate parts: nitrogen fixation, ammonification, nitrification, assimilation, and denitrification.

The Phosphorus Cycle

The phosphorus cycle is a sedimentary cycle (unlike carbon, oxygen, and nitrogen), the atmosphere is *not* a reservoir for phosphorous nor do microorganisms fix phosphorus as they do nitrogen. Phosphorus enters the biosphere almost entirely from the soil through absorption by plant roots. Weathering of rocks containing phosphate minerals, chiefly **apatite** [$Ca_5(PO_4)_3OH$], results in the relatively small pool of inorganic phosphorus available for organismal use. In most soils the major amount of phosphorus absorbed by plants comes from organic molecules that undergo decomposition releasing phosphorus in plant-available inorganic forms. The release of organically bound nutrients to plant-available forms is termed **mineralization**, a process important in the release to the soil of sulfur and nitrogen as well as phosphorus. Phosphorus is used by organisms in energy transfers (ATP, NAPD), as a component of nucleic acids (RNA, DNA), and as a structural element of membranes (phospholipids).

The phosphorus cycle has fewer compartments than the other major nutrient cycles and also has a significant "leak" of phosphorus back to its lithosphere reservoir from which it is returned to active cycling only after long intervals of geologic time. The combination of three factors makes phosphorus a nutrient of concern in most ecosystems:

- Most soils have only small amounts from the weathering of disjunctly distributed rocks.
- Phosphorus is more insoluble than other nutrients and less mobile, hence less phosphorus travels in the soil solution; roots generally must grow into a zone of phosphorus availability.
- Phosphorus that drains from the land to the ocean is used by organisms in the surface waters, but a considerable amount is lost to the sediments in the shells and bones of marine organisms and by precipitation and settling of phosphates.

Human activities alter the phosphorus cycle chiefly by adding more available phosphorus where little was available previously. Phosphate-containing detergents used in the 1960s were carried by sewage systems into rivers and lakes and were a boon to algae and microorganisms, which responded with exuberant flushes of growth. Widespread **eutrophication** resulted and detergent-makers were obliged to remove phosphates from their products. Eutrophication—the enrichment of fresh waters with nutrients—results in **blooms** of plankton and algae. Death of these organisms increases the populations of aerobic bacteria of decay which, in turn, deplete the dissolved oxygen in the waters, thereby killing fish and other aerobic organisms. The anaerobic microorganisms move in, and the fresh water becomes an unpleasant, smelly soup of decay.

Agricultural use of phosphate-containing fertilizers has increased as the acreage of farmlands has expanded over time. At first, guano (the dung of seabirds) was collected from deposits on seashore rocks and added to the fields, but demand by inland farmers for phosphate fertilizer stimulated the mining of phosphate deposits (ocean sediments of past geologic ages). These applications, too, wash out of the fields into the world's waters and also can cause eutrophication.

The Sulfur Cycle

Sulfur is one of the macronutrients required by plants and is obtained by them from the soil and from the atmosphere. It is present in proteins and gives a distinctive odor to many substances. It is also a component of the amino acid cysteine and is present in a large number of enzyme systems. Several groups of prokaryotes utilize and release sulfur.

The major reservoirs for sulfur in the global cycle are pyrite and gypsum (an evaporite of seawater) in the lithosphere and in seawater. Very little sulfur is present in living organisms, but within the marine muds and terrestrial bogs where organic matter accumulates under anaerobic conditions considerable amounts are present. The quantities cycled from these sources are small, but the distinctive rotten egg odor of H_2S is often prevalent in the air over such sites. Increasing amounts of atmospheric sulfur compounds are the direct result of human activities and are principal components of air pollution in industrial areas. Most are short-lived in the air and wash out forming acid rain downwind from the industrial sites. The sulfur cycle is no longer in balance.

The sulfur cycle resembles that of nitrogen in several respects, for example the short-term movements of both elements is through the atmosphere as a result of the metabolism of bacteria. The gases move rapidly in a closed cycle from the air to the soil and back. There are several subcycles: 1.) a long, deep time cycle of weathering, erosion, deposition, 2.) a predominately atmospheric cycle where bacteria metabolize dead organic matter and release sulfur to the atmosphere where it has a short residence time before being washed back to the soil by precipitation, 3.) a marine cycle where evaporation of sea spray releases sulfur to the atmosphere temporarily and from where it falls back into the sea, and 4.) a soil–plant cycle where organic sulfur in manure or other fertilizer is used to sustain soil microbes and plants.

CHAPTER 26
PLANT GEOGRAPHY

Distribution of Vegetation

Even the most cursory examination reveals that different kinds of plants grow in different kinds of places. The vegetation of the world is aligned latitudinally in broad bands circling the globe. As the climate changes from the equator northward and southward, so too, does the vegetation. The lush tropical rainforests of the equatorial band in the Northern Hemisphere give way to temperate deciduous forests, which in turn are replaced by coniferous forests that, at their northern limit, are replaced by treeless arctic tundra. More land is present north than south of the equator so the banding pattern is less pronounced in the Southern Hemisphere. High mountains on all continents also break the pattern.

The latitudinal bands of vegetation on the continents are replicated on a smaller scale by the altitudinal bands of vegetation on mountains.

Much more than temperature and precipitation changes environmentally with altitude, but the vegetational banding pattern remains: tundra on the tops of high mountains, coniferous forests in middle slopes, and deciduous forests at the base of mountains. The bands constrict and altitudinal limits become lower on mountains progressively northward from the equator. **Timberline,** which is the upper limit of tree growth and separates the alpine tundra from the coniferous forests, is at 10,000 feet in the southern Rocky Mountains, but at the Canadian border, is at 6,000 feet. Farther north it is lower still.

It is more than coincidental that vegetation and climate follow the same distributional patterns: Plants have **tolerance ranges**—ranges of environmental conditions—in which they can survive. Two environmental factors of great importance to plants are two that also determine climate—temperature and precipitation. There's more to the story than this, of course, but with available water and a moderate range of temperatures most plants will grow. As extremes are reached in both temperature and precipitation, specific kinds of plants disappear from the regional floras. In the tropics, there are over 40,000 species of vascular plants; in the forests of the southeastern United States, 5,000; in the Canadian arctic, about 425.

Figure 26-1 illustrates in diagrammatic form how vegetation types are related to temperature and precipitation. Tropical rain forests, for example, occur in the hottest, wettest regions of the world, deserts in the hottest, driest, tundra in the coldest, driest, and so forth.

Terrestrial Biomes of the World

The study of the distribution of plants is the science of plant geography. The units of vegetation the plant geographers study vary in size from the basic **community,** to groups of communities and their environment—**ecosystems**—to assemblages of ecosystems with distinctive vegetation and growth forms that extend over large geographic areas—**biomes.** Although these three terms each delineate portions of vegetation that occupy space, the terms imply no size per se. A plant community can be the lichens and mosses growing on the bark of a single tree trunk, an acre woodlot, or the plant communities of Yellowstone National Park. So, too, the size of ecosystems varies, although in this case energy flow and nutrient cycling dynamics determine the boundaries of the ecosystem; that is, more reactions occur *within* the system than *between* systems. Biomes are the largest units and are identified on continental, hemispheric, or worldwide scale. They are composed of similar **growth forms** and occur in broadly similar environments. The following paragraphs describe some commonly recognized terrestrial biomes.

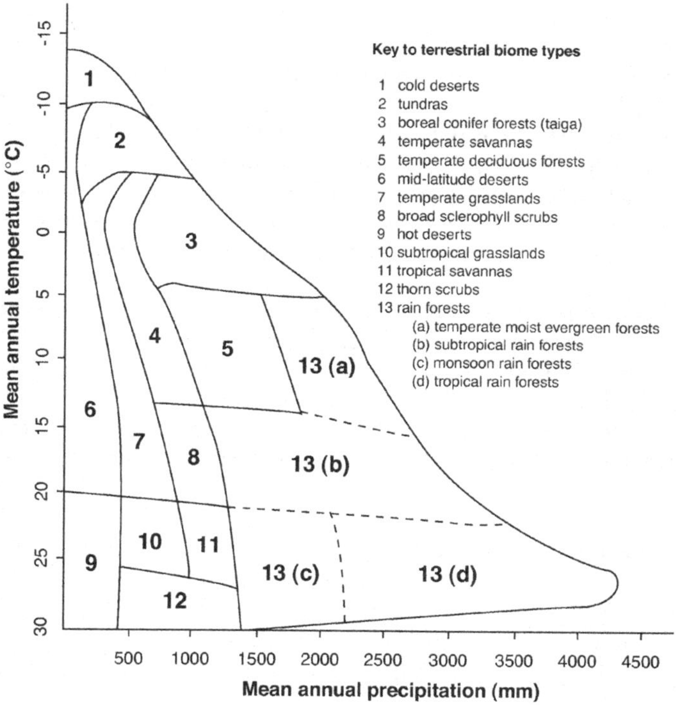

Figure 26-1

Arctic tundra occurs north of the tree line and principally north of the Arctic Circle in an area of low precipitation and little snow with cold average temperatures. Vegetation consists of small perennial herbs, low shrubs, creeping willows, and a variety of grasses, sedges, mosses and lichens.

Taiga is the Russian name for the coniferous forest that forms a wide belt between the tundra of the north and the temperate deciduous forest to the south. It is composed primarily of species of pine, spruce, and fir with an understory of ericaceous shrubs (heaths), mosses, and lichens. Over 65 percent is underlain by permafrost and thick peat deposits. Lakes and ponds are common and soils are poor.

Temperate deciduous forest is a mixture of broad-leaved deciduous trees (such as beech, maples, and oaks) together with species of perennial herbs. Seasons are pronounced with precipitation distributed evenly throughout the year. In North America this forest extends from the Atlantic coast westward to about 100° longitude.

Temperate moist evergreen forest, sometimes called "temperate rainforest," occurs in areas of heavy precipitation both north and south of the equator. In North America it is the forest of the northwestern Pacific Coast. Western hemlock, white cedar, coast redwood, spruce, and other trees attain large size with mosses and lichens festooning their branches and the whole forming a luxuriant, dense forest.

Temperate grasslands are the typical vegetation of the interior of continents where they cover thousands of acres. Although grasses are the dominant life form, trees occur along streams in *riparian* woodlands. Most of the highly fertile grasslands have been plowed and cropped and a majority of the remaining grasslands are managed as rangeland.

Warm deserts have hot summer temperatures with great diurnal temperature variations. Precipitation is slight and irregular. Shrubs, succulents, and annuals are common life forms in deserts worldwide. Productivity is low and limited by lack of moisture. Some deserts may have no rainfall for 10 to 15 years and plants survive on dew.

Cold deserts (and **polar deserts**) are dominated by shrubs. The growing season is condensed between cold winters and dry summers. In contrast to the grasslands, most of the photosynthate here goes to wood production rather than to digestible foods. *Polar deserts* lie closer to the pole in the high arctic and are colder and drier than the surrounding tundra and have only scattered patches of vegetation in protected spots.

Mediterranean (broad sclerophyll) scrub consists of shrubs and small trees with broad, hard, evergreen leaves called sclerophylls. Many of the shrubs produce secondary metabolites toxic to other plants, an example of allelopathy. Five areas occur worldwide on the southwestern coasts of continents in addition to the namesake area of the Mediterranean Basin. In California the type is called **chaparral.** All the areas have wet winters and warm, dry summers

Tropical savannas are grasslands with scattered trees and with three growing seasons, warm and rainy, cool and dry, hot and dry. There is no cold season. They are highly productive and in Africa, for example, support large populations of grazing and browsing hoofed herbivores and large carnivores.

Tropical rainforests have little climatic variation—no seasons and no cold nor dry period–and are either *tradewind* type (with steady, almost daily rains) or *equatorial* (with frequent, heavy thunderstorms). Trees are broad-leaved evergreens and are covered with lianas and epiphytes, which forms dense jungles. These are highly productive ecosystems with huge numbers of decomposers.

9 780764 585609

Made in the USA
Monee, IL
04 November 2026

67384038R00144